Lecture Notes in Mathematics

Edited by A. Dold and B. Eckmann

1008

Algebraic Geometry

Proceedings of the Third Midwest Algebraic
Geometry Conference held at the University of Michigan,
Ann Arbor, USA, November 14–15, 1981

Edited by I. Dolgachev

Springer-Verlag
Berlin Heidelberg New York Tokyo 1983

Editor

I. Dolgachev
Department of Mathematics, University of Michigan
Ann Arbor, Michigan 48109, USA

AMS Subject Classifications (1980): 14-06, 14 H 45, 14 J 25, 14 M 20, 14 H 10, 14 F 40, 14 J 17, 20 C 30, 14 E 20, 13 C 15

ISBN 3-540-12337-7 Springer-Verlag Berlin Heidelberg New York Tokyo
ISBN 0-387-12337-7 Springer-Verlag New York Heidelberg Berlin Tokyo

Printing and binding: Beltz Offsetdruck, Hemsbach/Bergstr.
2146/3140-543210

Preface

This volume represents the proceedings of the Third Midwest Algebraic Geometry Conference held at the University of Michigan on November 14-15, 1981.

The conference included five one hour lectures of the invited speakers from outside of Midwest (A. Bialynicki-Birula, D. Gieseker, N. Nygard and P. Slodowy) and three 45 minutes lectures of the geometers from Midwest (D. Burns, C. Huneke and T. Kambayashi). The research articles included in the present volume represent all the lectures except those by A. Bialynicki-Birula and P. Slodowy (which will be published elsewhere). I would like to thank P. Slodowy for allowing me to include his survey article on a topic closely related to the one of his actual lecture. Following the example of the Proceedings of the First Midwest Algebraic Geometry Conference (Lecture Notes in Mahtematics, vol. 862, 1981) I included in this volume an article of one of the participants (N. Goldstein) which was not presented orally.

It is my pleasure to thank all participants of the conference. I am also thankful to the National Science Foundation whose support made this meeting possible.

Igor V. Dolgachev

Table of Contents

ON THE GEOMETRY OF ELLIPTIC MODULAR SURFACES AND REPRESENTATIONS OF FINITE GROUPS

D. Burns*
DEPARTMENT OF MATHEMATICS
University of Michigan
Ann Arbor, Michigan 48109

Introduction. Let $\Gamma \subset Sl_2(\mathbb{Z})$ be a torsion-free subgroup of finite index. Associated to Γ is an elliptic surface $E(\Gamma)$ fibered over $X(\Gamma) := \overline{\Gamma \backslash \mathfrak{h}}$, where $\mathfrak{h} = \{\tau \in \mathbb{C} | \mathrm{Im}(\tau) > 0\}$, and Γ operates on $\mathfrak{h}$ in the usual manner. (cf [7] for the construction). If Γ is normal in Γ', $Sl_2(\mathbb{Z}) \supset \Gamma' \supset \Gamma$, then the finite group $G = \Gamma\backslash\Gamma'$ acts on $X(\Gamma)$ and $E(\Gamma)$, and $X(\Gamma)$, $E(\Gamma)$ have G-equivariant models in various $\mathbb{P}(V)$, for V a suitable representation space of G. Conversely, one may try to reconstruct $X(\Gamma)$, $E(\Gamma)$ and the presentation $1 \to \Gamma \to \Gamma' \to G \to 1$ from the geometry of the representation V (and its tensor representations). More specifically, one could try to find a G-invariant curve X in $\mathbb{P}(V)$, and construct a G-invariant surface E over X whose classifying map to the modular stack for elliptic curves could be used to intertwine G and $Sl_2(\mathbb{Z})$, or a subgroup of it. This latter was our initial motivation for the present paper, and remains largely unfulfilled, except for some elementary examples at the end of the paper. Nevertheless, we feel that some progress has been made in this direction, since several characteristic features of the geometry of $E(\Gamma)$ as an elliptic surface over $X(\Gamma)$ can be specified intrinsically.

The approach taken here is to consider the ruled surface over $X(\Gamma)$ obtained by quotienting $E(\Gamma)$ by the involution $\sigma(z) = -z$, where the sign refers to the group law in the fibers of $E(\Gamma)$. We can construct $E(\Gamma)$ over $X(\Gamma)$, if we can construct the ruled surface over $X(\Gamma)$, the branch curve and divisor class of the double covering.

Our first object in the paper is to construct a standard minimal model $S(\Gamma)$ for the ruled surface $\langle\sigma\rangle \backslash E(\Gamma)$. This is done for any $\Gamma \subset Sl_2(\mathbb{Z})$. The point is that this ruled surface can be specified over $X(\Gamma)$ purely geometrically, i.e., without reference to Γ.

Next, for Γ normal in $Sl_2(\mathbb{Z})$, (i.e., $\Gamma' = Sl_2(\mathbb{Z})$), we characterize the image B in $S(\Gamma)$ of the branch curve of $E(\Gamma)$ over

*A. P. Sloan Fellow. Partially supported by the National Science Foundation.

$\langle\sigma\rangle\backslash E(\Gamma)$ as the unique G-invariant divisor in its linear equivalence class. This can be generalized slightly, i.e., to $\Gamma' \neq Sl_2(\mathbb{Z})$, if $X(\Gamma')$ has genus 0 , and one knows enough about the cusps of $X(\Gamma')$.

Finally, for certain Γ , we calculate that there is only one G-invariant square-root of the line bundle $L(B)$ on $S(\Gamma)$.

Concretely, given <u>any</u> projective immersion $X(\Gamma) \to \mathbb{P}(V)$, $S(\Gamma)$ is the tangential ruled surface of $X(\Gamma)$: given $x \in X(\Gamma)$, the fiber $S(\Gamma)_x$ is the projective tangent line to $X(\Gamma)$ at x . The zero-section of $S(\Gamma)$ is simply the distinguished point x itself in the tangent line. If V is a G-representation space, and $X(\Gamma)$ is a G-invariant curve, to reconstruct the branch curve B , it suffices to choose the remaining three points (besides x) of $B \cap S(\Gamma)_x$ in a G-equivariant way. By the previous results, there is only one way to do this, and it is here that one uses most strongly the representation theory of G . In the examples we can calculate $(\Gamma = \Gamma(5), \Gamma(7), \Gamma_o(2) \cap \Gamma(5)$; $G = A_5$, $PSL(2;7)$, $A_5)$, one uses the decomposition of the tensor representations of V to construct G-equivariant families of hypersurfaces which cut out $B \cap S(\Gamma)_x$ as x moves on $X(\Gamma)$. The examples can be carried out in varying degrees of detail. The constructions carried out here should be compared with a construction of Naruki's [6] for the elliptic modular surface of level 5.

It is a pleasure to take this opportunity to thank I. Dolgachev for many long and fruitful discussions of the present topic. The difference between his contributions to this paper and those of a co-worker are negligible. Thanks are also due to R. Griess and R. Gunning, for useful remarks and references. The author also wishes to thank the Institute for Advanced Study and l'Université de Paris XI (Orsay) for their hospitality during the preparation of this paper.

§1. Standard Models for $\langle\sigma\rangle\backslash E(\Gamma)$.

As above $E(\Gamma)$ denotes the elliptic modular surface associated to $\Gamma \subset Sl_2(\mathbb{Z})$. If Γ is torsion free, then the singular fibers of $E(\Gamma)$ occur exactly over the cusps of $X(\Gamma)$. These singular fibers are of type I_b or I_b^* , $b \geq 1$, in Kodaira's notation, as the corresponding cusps are of the <u>first</u> or <u>second kind</u>, respectively, in the terminology of [7], for example. Recall that by construction, any $g \in Sl_2(\mathbb{Z})$ which normalizes Γ acts on $E(\Gamma)$ preserving the fibers and the 0-section. Let σ denote the involution of $E(\Gamma)$

induced by $-I \in Sl_2(\mathbb{Z})$. Over $\Gamma\backslash\mathfrak{h}$, σ fixes the zero section O_E and a smooth 3-fold section ϑ_E composed of all the non-trivial points of order two in the fibers. Thus, to construct non-singular minimal models of $\langle\sigma\rangle\backslash E(\Gamma)$, it is only necessary to study the behavior of σ near the singular fibers, as in [6]. We treat the cases I_b and I_b^* separately. (We call b the width of the cusp.)

i) I_b:

The cases b even or odd are slightly different. As in [6], we show the array of curves in a fiber of type I_b , and indicate the fixed points of σ :

b odd:

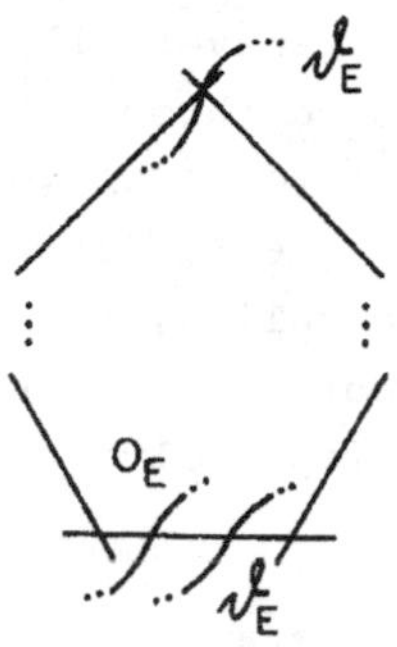

b even:

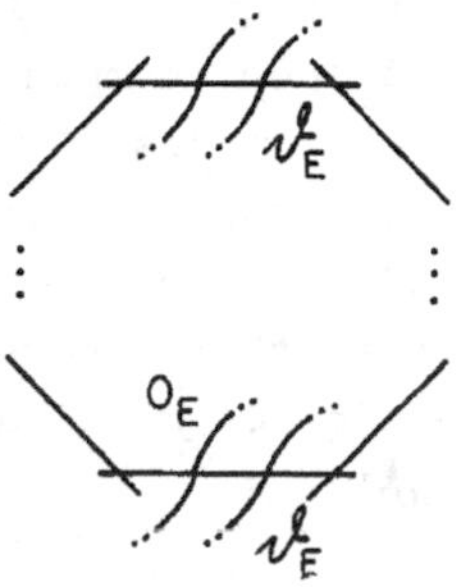

Each straight line represents a rational curve of self-intersection -2. In either case, σ has no isolated fixed points, and so $\langle\sigma\rangle\backslash E(\Gamma)$ is a smooth surface $\hat{S}(\Gamma)$ over $X(\Gamma)$, with special fiber over v :

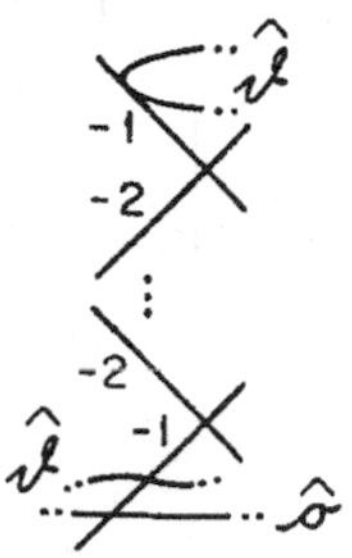

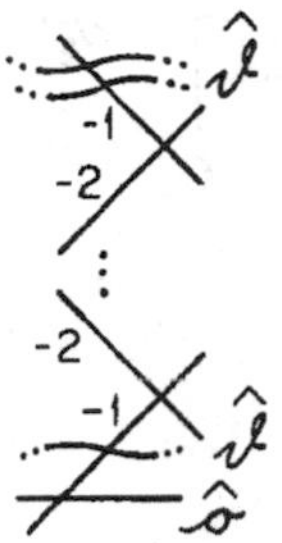

Here $\hat{\sigma}$ is the image in $\hat{S}(\Gamma)$ of O_E, and $\hat{\vartheta}$ is the image of ϑ_E. The components of the special fibers are again rational curves, with self-intersections as indicated. Note that the fibers $E(\Gamma)_v$ and $\hat{S}(\Gamma)_v$ are reduced.

The <u>canonical model</u> $S'(\Gamma)$ of $\langle\sigma\rangle\backslash E(\Gamma)$ is the $\mathbb{P}^1$-bundle over $X(\Gamma)$ obtained by blowing down the "top-most" exceptional curve of the first kind (ECFK) and continuing until the final situation below is realized:

The "lower" local component of ϑ' crosses $S'(\Gamma)_v$ transversally, and is the image of the local component of ϑ_E which passed through the same component of $E(\Gamma)_v$ as O_E. Note that $\sigma' \cdot \vartheta' = 0$. Also note that $S'(\Gamma)$ is canonical in the sense that any $g \in Sl_2(\mathbb{Z})$ which normalizes Γ will operate on $S'(\Gamma)$ preserving the ruling, σ' and ϑ'. Finally, note that reversing the above procedure amounts to the minimal succession of monoidal transforms needed to resolve the singularities of ϑ'.

ii) I_b^* :

Separate b even from b odd:

<u>b odd</u>:

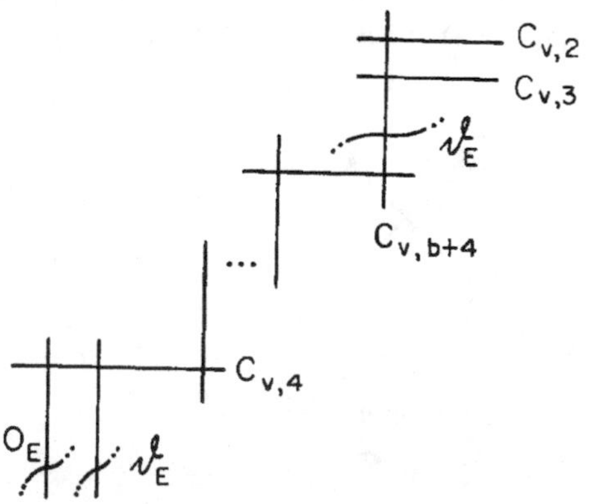

<u>b even</u>:

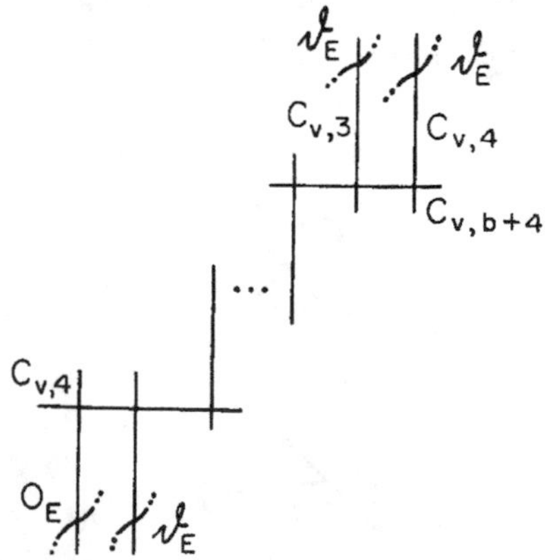

The irreducible components of E_v are labelled $C_{v,0},\ldots,C_{v,b+4}$, where $E_v = C_{v,0} + C_{v,1} + C_{v,2} + C_{v,3} + 2C_{v,4} + \ldots + 2C_{v,b+4}$. $0_E \cdot C_{v,0} = 1, \vartheta_E \cdot C_{v,1} = 1$. If b even, $C_{v,2} \cdot \vartheta_E = C_{r,3} \cdot \vartheta_E = 1$, and $C_{v,b+4} \cdot \vartheta_E = 0$, and σ fixes $C_{v,4}, C_{v,6},\ldots,C_{v,b+4}$ pointwise. If b odd, $C_{v,2} \cdot \vartheta_E = C_{v,3} \cdot \vartheta_E = 0$, $C_{v,b+4} \cdot \vartheta_E = 1$, and σ fixes $C_{v,4}, C_{v,6},\ldots,C_{v,b+3}$ pointwise. Thus, σ has no isolated fixed points, and the quotient $\langle\sigma\rangle\backslash E(\Gamma)$ is non-singular, and the fiber over v looks like:

b odd:

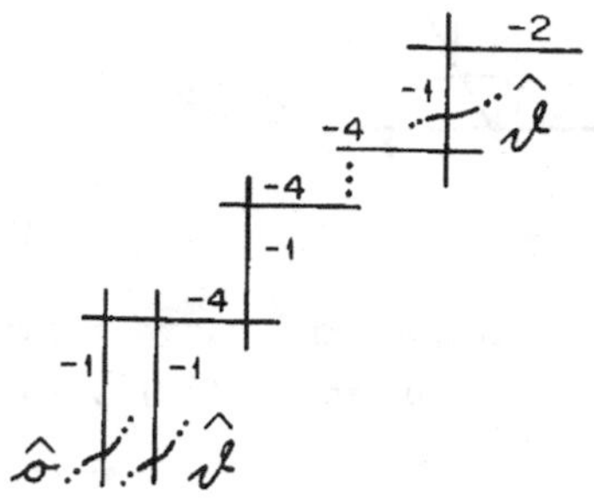

b even:

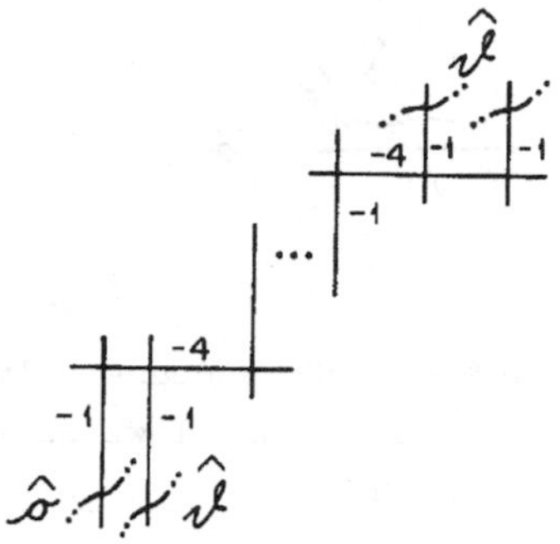

The self-intersections are marked. The components with self-intersection -4 have become simple components of the fiber $\hat{S}(\Gamma)_v$. All double components are ECFK. Blow these (disjoint) curves down, to get:

b odd:

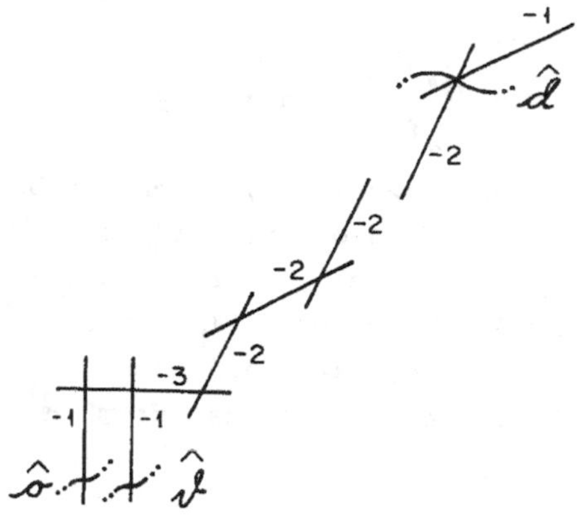

b even:

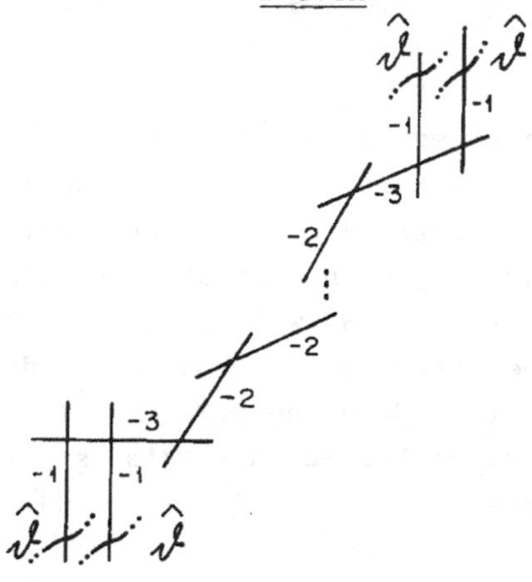

Next, blow-down again all the (disjoint) ECFK's, to get

b odd:

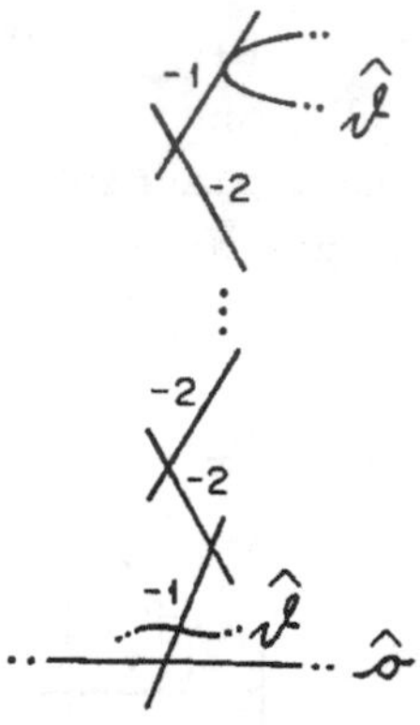

b even:

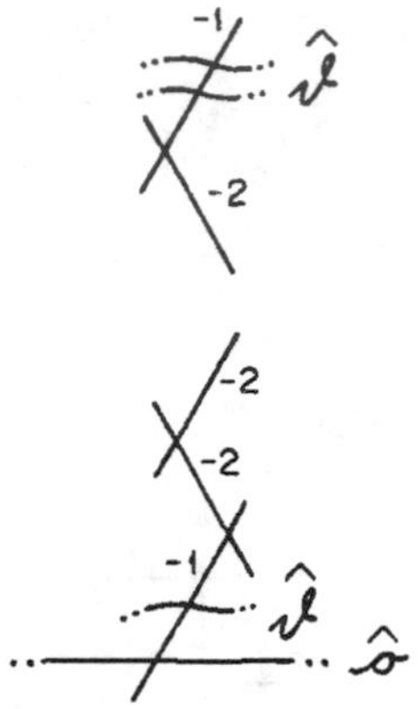

Finally, blow-down ECFK successively, starting at the curve farthest from $\hat{\sigma}$, and continue until we have a ruled surface $S'(\Gamma)$, with $S'(\Gamma)_v$ of the form:

b odd:

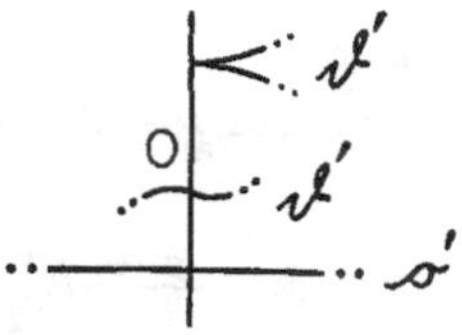

b even:

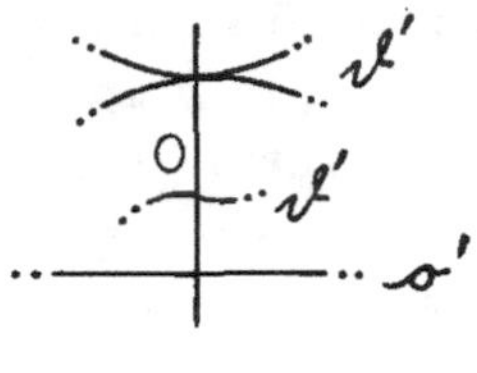

The local equations for the double points of ϑ' are again $y^2 = x^b$. As before, this $S'(\Gamma)$ is canonical, and $\sigma' \cdot \vartheta' = 0$.

Consider, next, another model $S(\Gamma)$ of $\langle\sigma\rangle \backslash E(\Gamma)$, which can be defined only when each singular fiber of $E(\Gamma)$ is of type I_b or I_b^*, with $b > 1$. We will call this the geometric model, for reasons that will be clear in §2. For this model, for fibers of type I_b, one first blows down the ECFK in $\hat{S}(\Gamma)$ which crosses $\hat{\sigma}$, and then proceeds to blow-down ECFK's, starting from the one farthest from $\hat{\sigma}$. One gets:

b odd:

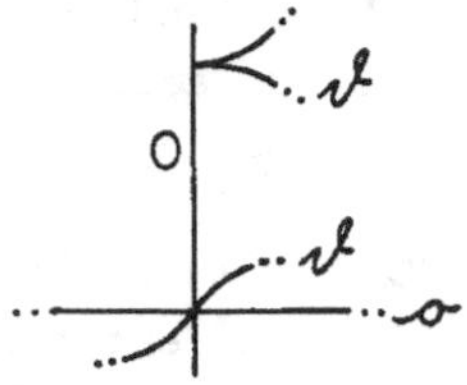

b even:

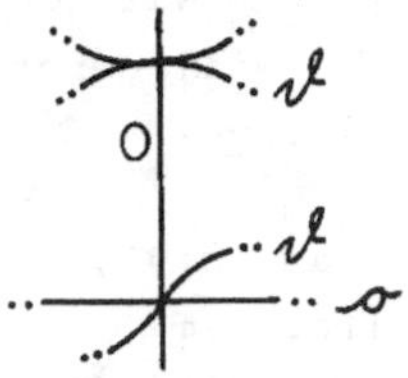

The local equation for ϑ near a point of intersection with $S(\Gamma)_v$ other than where it crosses σ simultaneously, is $y^2 = x^{b-2}$. (When $b = 2$, there are two such distinct transverse intersections.)

Similar adjustments can be made in the blowing down of $\hat{S}(\Gamma)$ near a cusp v with $E(\Gamma)_v$ of type I_b^*. The resulting intersection diagrams for σ, ϑ and $S(\Gamma)_v$ are exactly as for the corresponding I_b (same b) as above.

Note that for an elliptic modular surface with all cusp widths > 1, in the geometric ruled surface $S(\Gamma)$, the divisor ϑ intersects the section σ transversally, exactly at the cusps. Finally, note that the only automorphism of $S'(\Gamma)$ over $X(\Gamma)$, of finite order and fixing a cusp of width 1, is the identity.

§2. Intrinsic characterization of $S'(\Gamma)$, $S(\Gamma)$.

Recall that if S is a ruled surface over a curve X with projection π and a section σ, then we have a short exact sequence on S:

$$0 \to \mathcal{O}_S \to \mathcal{O}_S(\sigma) \to N_\sigma \to 0 \tag{2.1}$$

where N_σ is the normal sheaf to σ in S. Taking direct images via π, one has

$$0 \to \mathcal{O}_X \to \mathcal{E} \to N_\sigma \to 0 \tag{2.2}$$

and S is isomorphic to the projectivization $\mathbb{P}(E)$ of the rank 2 vector bundle E on X associated to the sheaf $\mathcal{E}$. $\mathcal{E}$ is determined by its extension class in $H^1(X,\mathcal{N}^*_{\vartheta})$.

For any variety, Θ_X will denote the tangent sheaf, and for $X = X(\Gamma)$ as above, κ will denote the divisor of cusps, each cusp counted with multiplicity 1.

Theorem 1': For Γ as in §1, $S = S'(\Gamma)$, then $\mathcal{N}_{\vartheta} \cong \Theta_X(-\kappa)$. The extension (2.2) splits. If Γ is normal in $Sl_2(\mathbb{Z})$, the extension splits in a unique $G = \Gamma\backslash Sl_2(\mathbb{Z})$ equivariant way.

Theorem 1: For Γ as above, with all cusp widths > 1 , and $S = S(\Gamma)$, then $\mathcal{N}_{\vartheta} \cong \Theta_X$. The extension (2.2) does not split.

Note that the statement of Theorem 1 determined E in (2.2), and hence $S(\Gamma)$, uniquely up to isomorphism, since $H^1(X(\Gamma),\mathcal{N}^*_{\vartheta}) = H^1(X(\Gamma),\Omega^1_{X(\Gamma)}) = \mathbb{C}$, and the extension class is well-determined modulo isomorphisms only up to multiplication by non-zero complex numbers: hence, the extension class is either 0 (split) or non-zero (unique non-split extension). The latter case arises quite naturally geometrically as the tangential ruled surface for any projective immersion of $X(\Gamma)$ in $\mathbb{P}^n$.

To calculate $\mathcal{N}_{\vartheta}$ as in theorems 1 and 1' , we pass to an auxiliary subgroup Γ_0 . Let Γ_o be a normal subgroup of finite index in $Sl_2(\mathbb{Z})$, with $\Gamma_0 \subset \Gamma$. If Γ is torsion-free, then $G_0 = \Gamma_0\backslash\Gamma$ acts on $S'(\Gamma_0)$ and $S'(\Gamma)$ is birational to $G_0\backslash S'(\Gamma_0)$, the map being biregular over $X(\Gamma)$ - {cusps} . We claim that this birational map from $G_0\backslash S'(\Gamma_0)$ to $S'(\Gamma)$ is biholomorphic. It suffices to check the behavior at the cusps.

Note, first of all, that the group G_0 has fixed points on $X(\Gamma_0)$ only at the cusps. If $\gamma \in G_0$ fixes the cusp $v \in X(\Gamma_0)$, it must act as the identity on $S'(\Gamma_0)_v$. Indeed, g preserves ϑ' and θ' in $S'(\Gamma_0)$, hence it fixes these points in $S'(\Gamma_0)_v$: where $S'(\Gamma_0)_v$ meets θ' , where it meets ϑ' transversally, and where it meets ϑ' with multiplicity 2. Since $S'(\Gamma_0)_v \cong \mathbb{P}^1$, g fixes $S'(\Gamma_0)_v$ pointwise. In particular, the quotient space $G_0\backslash S'(\Gamma_0)$ is non-singular, and a smooth ruled surface over $X(\Gamma)$. Let $S'' = G_0\backslash S'(\Gamma_0)$, and let θ'' , ϑ'' denote the images of the curves θ' and ϑ' in $S'(\Gamma_0)$ in S'' . Let $q : S'' \to S'(\Gamma)$, be birational and biregular except possibly over the cusps. At a cusp v of $X(\Gamma)$, trivialize $S'(\Gamma)$ by coordinates (z,ζ) where z is the coordinate on $X(\Gamma)$, with $\{z = 0\} = v$, and ζ an affine coordinate on $\mathbb{P}'$, so that $\mathcal{O}'$ is given locally by $\{\zeta = 0\}$. Assume, further that ζ is so normalized that the local component of $\mathcal{J}'$ which crosses $S'(\Gamma)_v$ transversally is given locally by $\zeta = \infty$, and that the double intersection of $\mathcal{J}'$ with $S'(\Gamma)_v$ occurs at $z = 0$, $\zeta = 1$. We can similarly find coordinates z, ζ'' on S'' so that $\{\zeta'' = 0\}$ is the image of $\mathcal{J}'(\Gamma_0)$ in S'', and the transverse local component of the image of $\mathcal{J}'(\Gamma_0)$ is given by $\zeta'' = \infty$, and the double intersection occurs at $\{z = 0, \zeta'' = q\}$. With these normalizations, the map q is given locally by

$$(z,\zeta'') \overset{q}{\longmapsto} (z, a(z)\cdot\zeta'') = (z,\zeta)$$

where $a(z)$ is a meromorphic function defined on a small disc about $z = 0$ and non-vanishing away from $z = 0$. Since q must send $\mathcal{J}''$ to $\mathcal{J}' = \mathcal{J}'(\Gamma)$, we see that $a(z) \to 1$ as $z \to 0$. Hence, q extends biholomorphically across $z = 0$, and $S'' \tilde{\to} S'(\Gamma)$. Let $N_{\mathcal{O}'}(\Gamma)$ be the normal bundle to $\mathcal{O}'$ in $S'(\Gamma)$, on $X(\Gamma)$, and $N_{\mathcal{O}'}(\Gamma_0)$ the analogous bundle in $X(\Gamma_0)$.

Similarly, if $p : X(\Gamma_0) \to X(\Gamma)$ denotes the quotient map, we see that $S'(\Gamma_0) \tilde{\to} p^*S'(\Gamma)$.

Next, note that if $\Theta_{X(\Gamma_0)}(-\kappa(\Gamma_0))$ and $\Theta_{X(\Gamma)}(-\kappa(\Gamma))$ are the sheaves for $S'(\Gamma_0)$ and $S'(\Gamma)$ in the statement of Theorem 1', and $T_{X(\Gamma_0)}(-\kappa(\Gamma_0))$ and $T_{X(\Gamma)}(-\kappa(\Gamma))$ the corresponding line bundles, then the differential dp_* establishes isomorphisms

$$T_{X(\Gamma_0)}(-\kappa(\Gamma_0)) \tilde{\to} p^*T_{X(\Gamma)}(-\kappa(\Gamma)), \quad \text{on } X(\Gamma_0) \tag{1}$$

$$G_0\backslash T_{X(\Gamma_0)}(-\kappa(\Gamma_0)) \tilde{\to} T_{X(\Gamma)}(-\kappa(\Gamma)), \quad \text{on } X(\Gamma). \tag{2}$$

For (2), note that the isotropy action of G_0 on $T_{X(\Gamma_0)}(-\kappa(\Gamma_0))$ at a cusp is trivial, so there is a well-defined holomorphic line bundle $G_0\backslash T_{X(\Gamma_0)}(-\kappa(\Gamma_0))$ on $X(\Gamma)$. Then the group G_0 acts on $N_{\mathcal{O}'}(\Gamma_0)$, again with trivial isotropy action over a fixed point v in $X(\Gamma_0)$ (since all $S'(\Gamma_0)_v$ will be fixed by the isotropy group of v in G_0), and we get isomorphisms:

$$N_{\mathcal{O}'}(\Gamma_0) \tilde{\to} p^* N_{\mathcal{O}'}(\Gamma) \ , \quad \text{on} \quad X(\Gamma_0) \ , \text{ and}$$

$$G_0 \backslash N_{\mathcal{O}'}(\Gamma_0) \tilde{\to} N_{\mathcal{O}'}(\Gamma) \ , \text{ on } \ X(\Gamma) \ .$$

As a result of this sequence of observations, we claim that if we can find one Γ_1 normal, torsion-free, of finite index in $Sl_2(\mathbb{Z})$, with $N_{\mathcal{O}'}(\Gamma_1) \tilde{\to} T_{X(\Gamma_1)}(-\kappa(\Gamma_1))$ in a $G_1 = \Gamma_1 \backslash Sl_2(\mathbb{Z})$ - equivariant way, then we have an isomorphism $N_{\mathcal{O}'}(\Gamma) \tilde{\to} T_{X(\Gamma)}(-\kappa(\Gamma))$, for any Γ torsion free, of finite index in $Sl_2(\mathbb{Z})$, which is natural in the sense that the isomorphism is equivariant for the action of the normalizer of Γ in $Sl_2(\mathbb{Z})$ on $N_{\mathcal{O}'}(\Gamma)$ and $T_{X(\Gamma)}(-\kappa(\Gamma))$. Indeed, one only has to choose Γ_0 above sufficiently small that it is contained normally in both Γ and Γ_1 . The above observations say the desired isomorphism "pulls back" from Γ_1 to Γ_0 , and "descends" from Γ_0 to Γ . However, the work of Naruki shows Theorem 1' is true for $\Gamma_1 = \Gamma(5)$, the principal congruence subgroup of level 5. (This follows from [6], §§2, 3 ; it also follows from the construction of $E(\Gamma(5))$ below in §4, and the remarks below on the relationship of $S'(\Gamma)$ and $S(\Gamma)$.) Thus, we know $N_{\mathcal{O}'}(\Gamma) \tilde{\to} \Theta_{X(\Gamma)}(-\kappa(\Gamma))$, for all $\Gamma \subset Sl_2(\mathbb{Z})$ as in Theorem 1'.

Consider now the extension (2.2):

$$0 \to \mathcal{O}_{X(\Gamma)} \to \mathcal{E} \to \Theta_{X(\Gamma)}(-\kappa(\Gamma)) \to 0 \ . \tag{2.3}$$

The extension class lies in $H^1(X(\Gamma)\ , \Omega^1_{X(\Gamma)}(\kappa(\Gamma)))$, which is zero, since $\deg \kappa(\Gamma)$ is always positive. Thus $\mathcal{E}$ splits. To complete the proof of Theorem 1', consider more closely the case when $\Gamma \subset Sl_2(\mathbb{Z})$ is normal. Tensor (2.3) with $\Omega^1_{X(\Gamma)}(\kappa(\Gamma))$, and take cohomology to get:

$$\begin{aligned} 0 \to H^0(X(\Gamma)\ , \Omega^1_{X(\Gamma)}(\kappa(\Gamma))) &\hookrightarrow H^0(X(\Gamma)\ , \mathcal{E}(K_{X(\Gamma)}+\kappa(\Gamma))) \\ &\hookrightarrow H^0(X(\Gamma), \mathcal{O}_{X(\Gamma)}) \to 0 \ . \end{aligned} \tag{2.4}$$

$G = \Gamma \backslash Sl_2(\mathbb{Z})$ operates equivariantly on (2.4). A G-invariant splitting of (2.3) is a G-invariant element $\phi \in H^0(X(\Gamma), \mathcal{E}(K_{X(\Gamma)}+\kappa(\Gamma)))$

restricting to $1 \in H^0(X(\Gamma), \mathcal{O}_{X(\Gamma)})$. It suffices to show that the invariants $\{H^0(X(\Gamma), \Omega^1_{X(\Gamma)}(\kappa(\Gamma)))\}^G = 0$. But this follows from the following observations:

(1) such an invariant would be a meromorphic differential ω with log-poles at the cusps, invariant under G.

(2) the cusps in $X(\Gamma)$ are one orbit under the action of G

(3) $G\backslash X(\Gamma) = X(Sl_2(\mathbb{Z})) = \mathbb{P}^1$

(4) (1) - (3) taken together imply ω descends to a differential $\tilde{\omega}$ with one log-pole on $\mathbb{P}^1$, which is necessarily 0. Hence $\omega = 0$, completing the proof of Theorem 1'.

The proof of Theorem 1 is very similar. Let $\Gamma \subset Sl_2(\mathbb{Z})$ be as in Theorem 1. To calculate the normal bundle of $\mathcal{O}(\Gamma)$ in $S(\Gamma)$, we compare it with the canonical model $S'(\Gamma)$. It is easy to see in the constructions of §1 that to pass from $S'(\Gamma)$ to $S(\Gamma)$, we blow up, for each cusp v of Γ, the point on $S'(\Gamma)_v$ where ϑ has its double intersection with $S'(\Gamma)_v$, then blow down the proper transform of the old fiber $S'(\Gamma)_v$. The section $\mathcal{O}$ is the birational transform of $\mathcal{O}'$, and its normal bundle is the normal bundle of $\mathcal{O}'$, namely $T_{X(\Gamma)}(-\kappa(\Gamma))$, by Theorem 1', twisted by exactly the divisor $\kappa(\Gamma)$, since we've blown down one ECFK at every cusp on $\mathcal{O}'$. Thus, $T_{X(\Gamma)}$ is the normal bundle.

As noted above, to calculate the extension (2.2) for $S(\Gamma)$, it now suffices to check that (2.2) doesn't split, or, equivalently, that $S(\Gamma)$ doesn't admit a section which doesn't intersect $\mathcal{O}$. To do this, we pass to a subgroup $\Gamma_0 \subset \Gamma$, as in the proof of Theorem 1'. With notation as above, we have a birational map $S(\Gamma_0) \to q^*(S(\Gamma))$, which is biholomorphic except over the cusps in $X(\Gamma_0)$. Let v_0 be a cusp of Γ_0, and $v = q(v_0)$ the corresponding cusp of Γ. Near $S(\Gamma)_v$ and $q^*S(\Gamma)_{v_0}$ we have the following diagrams:

In suitable local coordinates on $X(\Gamma)$ and $S(\Gamma)$, the transverse branch of ϑ is given by $x = y$, ϑ is given by $y = 0$, and, with

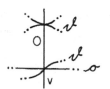

another coordinate $\bar{y}$, ϑ near its other intersections with $S(\Gamma)_v$ is given by $\bar{y}^{-2} = x^{b-2}$ (interpreted as two disjoint branches when $b = 2$) , b = width of the cusp v . If the width of the cusp v_0 is kb , k an integer, choose a local coordinate ξ on $X(\Gamma_0)$ at v_0 so that q is given by $x = \xi^k$. Setting $\eta = q^*y$ near v_0 in $q^*(S(\Gamma))$, then $q^{-1}(\sigma)$ is given by $\eta = 0$, and $q^{-1}(\vartheta)$ near v_0 is given by $\eta = \xi^k$, and near the other intersections of $q^{-1}(\vartheta)$ and $q^*(S(\Gamma))_{v_0}$, $q^{-1}(\sigma)$ is given by $\eta^{-2} = x^{bk-2k}$. Blow up v_0 (k-1)-times so that the proper transform of $q^{-1}(\sigma)$ intersects the proper transform of $q^{-1}(\sigma)$ transversally at v_0 . Then blow down successively (k-1) - ECFK in the new fiber over v_0 (cf. the picture below).

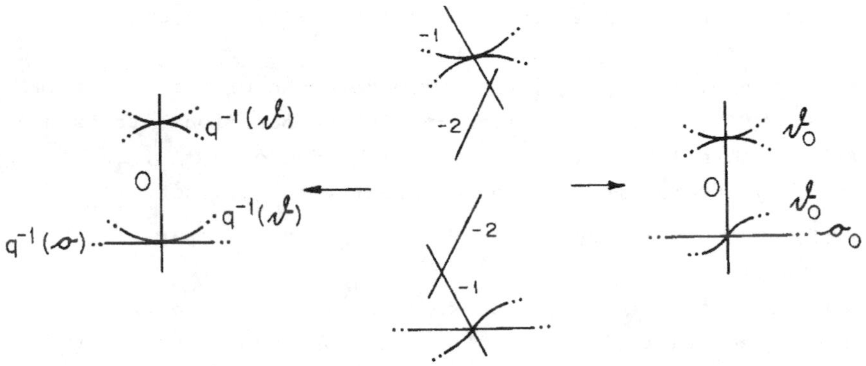

(2.5)

Call this new model S_0 . We have a birational map $S(\Gamma_0) \to S_0$, biholomorphic except over the cusps. Note that ϑ_0 and σ_0 have the same local equations near $(S_0)_{v_0}$ as $\vartheta(\Gamma_0)$ and $\sigma(\Gamma_0)$. An argument very similar to the one used earlier proving Theorem 1' now shows that $S(\Gamma_0) \to S_0$ is in fact biholomorphic.

Now suppose $S(\Gamma)$ had a section Σ disjoint from $\sigma(\Gamma)$. Then it is easy to see from (2.5) that the proper transform Σ_0 of $q^{-1}(\Sigma)$ in S_0 would give a section of S_0 disjoint from σ_0 , and hence the sequence (2.2) for $S(\Gamma_0)$ would split.

Conversely, suppose (2.2) were to split for $S(\Gamma_0)$. We would have

$$0 \to \mathcal{O}_{X(\Gamma_0)} \to E \to \Theta_{X(\Gamma_0)} \to 0 \tag{2.6}$$

Tensoring with $\Theta^*_{X(\Gamma_0)} = \Omega^1_{X(\Gamma_0)}$, we would get

$$0 \to \Omega^1_{X(\Gamma_0)} \to \mathcal{E}(K_{X(\Gamma_0)}) \to \mathcal{O}_{X(\Gamma_0)} \to 0$$

split exact, hence:

$$0 \to H^0(X(\Gamma_0),\Omega^1_{X(\Gamma_0)}) \to H^0(X(\Gamma_0),\mathcal{E}(K_{X(\Gamma_0)})) \to H^0(X(\Gamma_0),\mathcal{O}_{X(\Gamma_0)}) \to 0$$

By complete reducibility in characteristic 0 , there is a G_0-invariant section in $H^0(X(\Gamma_0),\mathcal{E}(K_{X(\Gamma_0)}))$ restricting to $1 \in H^0(X(\Gamma_0),\mathcal{O}_{X(\Gamma_0)})$. Thus, if (2.6) splits for $S(\Gamma_0)$, then it splits G_0-equivariantly, and without loss of generality, we may assume Σ_0 in $S(\Gamma_0)$ is G_0-invariant. Then $G_0 \backslash \Sigma_0$ gives a section of $S(\Gamma)$, disjoint from $\mathcal{O}$: this disjointness requires verification only over the cusps of $X(\Gamma)$. This follows by reversing the procedure in (2.5) above, once we note that Σ_0 has to pass through the double point of $\vartheta(\Gamma_0)$ $(=\vartheta_0$ in (2.5)) on $S(\Gamma_0)_{v_0}$. Indeed, $\Sigma_0 \cap S(\Gamma_0)$ is invariant under the (non-trivial) isotropy group G_{v_0} of v_0 in G_0 , as are the two points of intersection of $\vartheta(\Gamma_0)$ and $S(\Gamma_0)_{v_0}$. We claim that this isotropy group acts non-trivially on $S(\Gamma_0)_{v_0}$. If so, then $\Sigma_0 \cap S(\Gamma_0)_{v_0}$ must be one of the two points of intersection of $\vartheta(\Gamma_0)$ and $S(\Gamma_0)_{v_0}$. Since it can't be v_0 , it must be the double point.

To see that G_{v_0} acts non-trivially on $S(\Gamma_0)_{v_0}$, note that at $v_0 \in S(\Gamma_0)_{v_0}$, G_{v_0} preserves three mutually transverse non-singular curves: $\mathcal{O}(\Gamma_0)$, $S(\Gamma_0)_{v_0}$ and the local branch of $\vartheta(\Gamma_0)$. Hence the representation of G_{v_0} on the tangent space to $S(\Gamma_0)$ at v_0 is by scalars ("If a 2×2 matrix has 3 distinct eigenvectors, it's a scalar multiple of the identity"). If G_{v_0} were to act trivially on $S(\Gamma_0)_{v_0}$, this representation would be trivial. Since G_{v_0} is a finite group, and we are in characteristic zero, this last would imply G_{v_0} acts trivially on $S(\Gamma_0)$ and on $X(\Gamma_0)$. But G_0 clearly acts effectively on $X(\Gamma_0)$ and G_{v_0} must have order ≥ 2 , by our choice of Γ_0 . This is a contradiction.

The preceeding arguments show, as in the proof of Theorem 1', that if we find any $S(\Gamma_1)$ for which (2.2) doesn't split, then (2.2) splits for no $S(\Gamma)$. In §4 below we construct $E(\Gamma)$, $S(\Gamma)$ for $\Gamma = \Gamma(5)$. Then $S(\Gamma(5))$ does not split. This is also proved by Naruki [6], §2. (To obtain Theorem 1' from [6] or §4 below, one passes from $S(\Gamma(5))$ to $S'(\Gamma(5))$ as above. Thus the logic is to show Theorem 1 for $\Gamma = \Gamma(5)$. This implies Theorem 1' for $\Gamma(5)$, which implies Theorem 1' for all Γ . Theorem 1' for all Γ and Theorem 1 for $\Gamma = \Gamma(5)$ are then used to prove Theorem 1 for general Γ .)

We conclude this § with the remark that the surface $S(\Gamma)$ is "more intrinsic" then $S'(\Gamma)$ in the sense that Theorem 1 tells how to construct $S(\Gamma)$ without reference to the divisor of cusps. In principle, then, one could try to prove a projective curve X is of the form $X(\Gamma)$ by starting from its non-trivial tangential ruled surface $S(X)$ with a section with normal bundle = T_X , and trying to reconstruct a suitable elliptic surface which double covers this surface. In practice this proves to be rather difficult, and we have only been able to calculate very simple examples in §4. In order to pass from the ruled surface $S(X)$ to an elliptic surface, one has first to construct the analogue of the divisor ϑ , the non-trivial component of the branch locus of σ in $S(\Gamma)$.

§3. Determining $\vartheta \subset S(\Gamma)$.

Let $\Gamma \subset Sl_2(\mathbb{Z})$ be as in §1, admitting a geometric model $S(\Gamma)$ (i.e., all $b \geq 2$) . In this § we will denote $S(\Gamma)$ simply by S , and $X(\Gamma)$ by X . The divisor $\vartheta \subset S$ is in the linear system of the divisor class $D = 3\Theta + 3K_X + \kappa$, where K_X and κ denote, respectively, the canonical divisor and the divisor of cusps (each with multiplicity 1) on X , pulled back to S . We know ϑ crosses Θ simply exactly at the cusps. It is not clear what other geometric conditions specify ϑ uniquely in its divisor class. However, if Γ is normal in $Sl_2(\mathbb{Z})$, and $G = \Gamma\backslash Sl_2(\mathbb{Z})$, then we know G leaves ϑ invariant. This characterizes ϑ .

Theorem 2: If Γ is normal in $Sl_2(\mathbb{Z})$, and has all cusps of width $b \geq 2$, then ϑ is the unique G-invariant effective divisor linearly equivalent to $D = 3\Theta + 3K_X + \kappa$.

Proof: There is an action of G on $\mathcal{O}_S(D)$, and the statement amounts to the claim $\dim H^o(S,\mathcal{O}_S(D))^G = 1$. To compute this, we restrict to $X = \vartheta \subset S$ three times:

$$0 \to \mathcal{O}_S(D-\vartheta) \to \mathcal{O}_S(D) \to \mathcal{O}_X(\kappa) \to 0 \tag{3.1}$$

$$0 \to \mathcal{O}_S(D-2\vartheta) \to \mathcal{O}_S(D-\vartheta) \to \Omega^1_X(\kappa) \to 0 \tag{3.2}$$

$$0 \to \mathcal{O}_S(3K_X+\kappa) \to \mathcal{O}_S(D-2\vartheta) \to \Omega^{\otimes 2}_X(\kappa) \to 0 \tag{3.3}$$

We've used Theorem 1 to say that $\mathcal{O}_S(\vartheta)$ restricts to Θ_X on ϑ . The last term in (3.3) is the sheaf of quadratic differentials with no worse than simple poles at the cusps.

Since we already know that one G-invariant section of $\mathcal{O}_S(D)$ defines ϑ , passing to cohomology in (3.1), (3.2), (3.3) shows that it suffices to prove that

$$\dim H^o(X,\mathcal{O}_X(\kappa))^G = 1 \tag{3.4}$$

$$\dim H^o(X,\Omega^{\otimes i}_X(\kappa)) = 0 \ , \ i = 1 \ , \ 2 \ , \ 3 \ . \tag{3.5}$$

To see this, consider

$$0 \to \mathcal{O}_X \xrightarrow{\sigma} \mathcal{O}_X(\kappa) \to \bigoplus_{v\in\kappa} \mathbb{C}_v \to 0 \ , \text{ and} \tag{3.6}$$

$$0 \to H^o(X,\mathcal{O}_X) \to H^o(X,\mathcal{O}_X(\kappa)) \to \bigoplus_{v\in\kappa} \mathbb{C}_v \ . \tag{3.7}$$

Here σ is the unique (G-invariant) section of $\mathcal{O}_X(\kappa)$ vanishing simply at the cusps. At a cusp $v_0 \in X$, the isotropy group $G_{v_0} \subset G$ of v_0 (which is non-trivial, since the cusp width $b \geq 2$) acts on $\mathbb{C}_{v_0}$ through a character ρ_{v_0} . In fact,

$\mathbb{C}_{v_0}$ = {germs of functions meromorphic near v_0 with a pole of order ≤ 1} modulo {germs of functions holomorphic at v_0}

and the action of $g \in G_{v_0}$ sends the class of the meromorphic function f at v_0 to $(g^{-1})^*f$. If in a suitable local coordinate z, $g : z \to \zeta_\ell z$, $\zeta_\ell = e^{2\pi i/\ell}$, then q acts by multiplication by ζ_ℓ on $\mathbb{C}_{v_0}$. Hence, the character ρ_{v_0} is non-trivial on G_{v_0}. Note also that, as a G representation space, the third term of (3.7) is just ρ_{v_0} induced up to G:

$$\bigoplus_{v \in \kappa} \mathbb{C}_v \simeq \mathrm{Ind}_{G_{v_0}}^{G} (\rho_{v_0}) \ .$$

By reciprocity,

$$\left(\bigoplus_{v \in \kappa} \mathbb{C}_v \right)^G = \{0\} \ .$$

This proves (3.4).

Since (3.5) is similar to an argument in §2, we treat only $i = 3$, the cases $i = 1, 2$ being easier. An invariant cubic differential ω on X with simple poles at the cusps descends to a meromorphic cubic differential $\tilde{\omega}$ on $G\backslash X = \mathbb{P}^1$, regular except at the three points $0, 1, \infty$ over which $X \to G\backslash X$ ramifies. This map has ramification only at the orbit of cusps (which goes to ∞), and along the orbits of elliptic fixed points of G of order 2 and 3, going to $0, 1$, respectively in $\mathbb{P}^1$, in the usual normalization. A simple local calculation shows that $\tilde{\omega}$ can have at most a double pole at ∞ and 1, and at most a simple pole at 0. Hence, $\tilde{\omega}$ is a section of $H^0(\mathbb{P}', \Omega^{\otimes 3}_{\mathbb{P}^1}(5)) = 0$. Thus, $\omega = 0$.

§4. Examples of constructions of elliptic modular surfaces.

In this section we consider three examples of a curve X in $\mathbb{P}^2$ invariant by a finite simple group G acting on $\mathbb{P}^2$. The curves we treat are well-known to be modular curves (cf. [1], for example). We try here to construct the corresponding elliptic modular surfaces over these curves directly in terms of the geometry and the representations involved, using the characterization results of the previous sections to motivate the choices made in the constructions.

There are only three finite simple groups G embedded in

$PGL(3,\mathbb{C})$: A_5, $PSL(2,7)$ and A_6 (cf. [5], for example). The invariant polynomials for these three actions were calculated by Klein, and those of lowest degree are relatively easy to calculate by character formulas. We will consider the following curves:

1) A_5 acts on an invariant (non-singular) quadric X, the icosahedron, which is $X(5)$, anti-canonically embedded.

2) $PSL(2,7)$ acts on an invariant (non-singular) quartic, the Klein quartic, which is $X(7)$, canonically embedded.

3) A_5 acts on a unique invariant sextic X with six nodes, Bring's curve (or Klein sextic in [6]), which is isomorphic to $\overline{\Gamma_0(2) \cap \Gamma(5)\backslash\mathfrak{h}}$.

As the complexity of the above examples increases, our constructions become less complete. Thus, for the icosahedron, we can construct the elliptic modular surface and all of its sections directly. For Klein's quartic, we can construct the elliptic surface, but the sections are too complicated to extract using the present methods. For Bring's curve, we can also complete the construction of the elliptic surface. It is interesting to see the effect on the construction of $\mathcal{V}$ of the existence of the lower order G-invariant, namely the quadric in 1).

Example 1: The Icosahedron.

$G = A_5$ acts irreducibly on $V = \mathbb{C}^3$ with an invariant (non-singular) quadric $X \subset \mathbb{P}(V) = \mathbb{P}^2$. We seek, as in the introduction, a G-equivariant family of cubic curves in $\mathbb{P}(V)$ to cut out the triple section $\mathcal{V}$ on the tangent lines to X. Considering the tangential ruled surface $S(X)$ as embedded in $\mathbb{P}(V) \times \mathbb{P}(V)$, let us consider G-invariant, bihomogeneous forms $P(z,\zeta)$ in $S^3(V^*) \otimes S^3(V^*)$ with the diagonal action. First note that, as G-representation,

$$S^3(V^*) \cong V \oplus V' \oplus V_4$$

where V' is the other (conjugate over $\mathbb{Q}$) irreducible 3-dimensional representation of A_5, and V_4 is the unique 4-dimensional irreducible representation of A_5. V_4 has a G-invariant quadratic form in $V_4^* \otimes V_4^* \simeq V_4 \otimes V_4$. We take this as our $P(z,\zeta) \in V_4 \otimes V_4 \subset S^3(V^*) \otimes S^3(V^*)$. The intersection of the hypersurface $\{P(z,\zeta) = 0\}$ and $S(X)$ in $\mathbb{P}(V) \times \mathbb{P}(V)$ determines a G-invariant triple section ϑ of $S(X)$. We'll next describe this more geometrically.

More concretely, $V_4 \subset S^3(V^*)$ is the space of cubics vanishing at the six "fundamental points" of $\mathbb{P}(V)$ which can be characterized as the unique G-orbit of (minimal) cardinality 6 in $\mathbb{P}(V)$. This linear system defines a rational map $\Phi : \mathbb{P}(V) \to \mathbb{P}(V_4^*)$ whose image is the unique (non-singular) G-invariant cubic C in $\mathbb{P}(V_4^*)$. The quadric X does not pass through the fundamental points, since the minimal G-orbit on X has cardinality 12. $\Phi(X)$ is a curve of degree 6 on C. In $\mathbb{P}^3 = \mathbb{P}(V_4^*)$, one associates to each $w \in \mathbb{P}^3$ the polar plane H_w with respect to the G-invariant quadric Q in $\mathbb{P}^3$. This, in turn cuts out a plane section of C which projects under Φ^{-1} to a cubic curve C_w in $\mathbb{P}^2$ passing through the six fundamental points. Thus, to $z \in X$ we've associated the cubic $C_{\Phi(z)}$, defining a triple of points on the tangent L_z to X through z. The so-defined triple section ϑ of $S(X)$ crosses σ at those $z \in X$ for which $\Phi(z)$ is contained in its own Q-polar in $\mathbb{P}^3$. This is exactly when $\Phi(z)$ is contained in the intersection Y of Q and C in $\mathbb{P}^3$. Under Φ^{-1}, Y projects to Bring's sextic curve in $\mathbb{P}^2$, and $\Phi^{-1}(Y)$ intersects X at 12 points counted with multiplicity. By G-invariance, these must be the orbit of fixed points of degree 5 on X.

Let κ denote the divisor of fixed points of degree 5 on X. On $S(X)$, it's clear that the divisor ϑ is linearly equivalent to $3\sigma + 3K_X + \kappa$. For $v \in \kappa$, and $\gamma \in G$ of order 5 fixing v, γ has two fixed points we'll denote by $0_v = S(X)_v \cap \sigma$ and ∞_v. We claim the following facts about ϑ as just constructed:

i) ϑ is irreducible

ii) for $v \in \kappa$, ϑ has a cusp at ∞_v

iii) for $z \notin \kappa$, ϑ intersects $S(X)_z$ transversally in three distinct points.

We will not carry out the details here. The proof follows from analyzing numerical invariants, comparing the genus formula for $\vartheta \subset S(X)$ and Hurwitz's formula for the projection $\pi: \vartheta \to X$. To resolve the singularities of $\vartheta \cup \mathcal{O}$, then, it suffices to blow-up $S(X)$ at O_v and ∞_v, for each $v \in \kappa$. The proper transforms $\hat{\vartheta}$ and $\hat{\mathcal{O}}$ on the blow-up $\hat{S}(X)$ are disjoint and non-singular. The fiber $\hat{S}(X)_v$ can be labelled as follows:

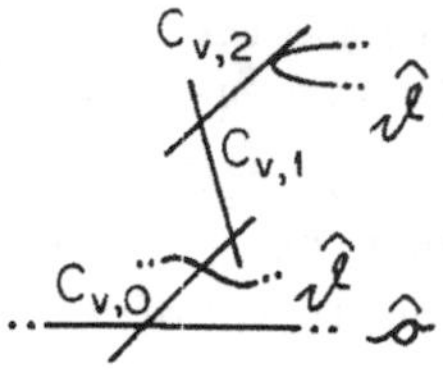

As divisior class on $\hat{S}(X)$,

$$B = \hat{\mathcal{O}} + \hat{\vartheta} \sim 4\hat{\mathcal{O}} + 3K_X + \kappa - 2 \cdot \sum_{v \in \kappa} C_{v,1} - 4 \cdot \sum_{v \in \kappa} C_{v,2} .$$

Since $\kappa \sim -6K_X$ on X, define

$$D = \tfrac{1}{2}B = 2\hat{\mathcal{O}} - 2K_X - \sum_{v \in \kappa} C_{v,1} - 2 \cdot \sum_{v \in \kappa} C_{v,2} + \tfrac{1}{2} K_X ,$$

where $\frac{1}{2} K_X$ is the unique (therefore G-invariant) θ-characteristic of X_0. Note that $\frac{1}{2} K_X$ is given geometrically as $-P$, P any point on X. The divisor class D is G-invariant, and the universal cover $\hat{G} = SL(2,5)$ acts on $O_X(D)$. Define $E(X)$ as the double cover of $\hat{S}(X)$, branched over B, corresponding to the divisor class D. $\hat{G}$ acts on $E(X)$. It is easy to see from the above that $E(X)$ has only I_5-singular fibers over $v \in \kappa$.

To construct sections of $E(X)$, consider again the six fundamenta points of G in $\mathbb{P}^2$. There are six conjugacy classes in G of cyclic-subgroups $\langle\gamma_5\rangle$ of order 5. The normalizer $N(\gamma_5)$ of $\langle\gamma_5\rangle$ in G is a dihedral group of order 10, and these six subgroups are the six isotropy groups in G of the fundamental points. Furthermore, each $\langle\gamma_5\rangle$ has two fixed points on X. We denote these points by $\kappa(\gamma_5)$. $N(\gamma_5)$ has two orbits in κ; it acts simply transitively on the complement of $\kappa(\gamma_5)$. If v_1 and v_2 are the two points of $\kappa(\gamma_5)$, the tangents to X at v_1 and v_2 intersect at the fundamental point P fixed by $N(\gamma_5)$. Fixing $\langle\gamma_5\rangle \subset G$, there is a unique quadric $Q(\gamma_5)$ in $\mathbb{P}^2$ passing through the 5 fundamental points not fixed by γ_5 . $Q(\gamma_5)$ is non-singular and invariant by $N(\gamma_5)$. It intersects X in 4 points, counted with multiplicity, and since the intersection is $N(\gamma_5)$-invariant , $Q(\gamma_5)$ is tangent to X at each of v_1 and v_2 in $\kappa(\gamma_5)$, and doesn't meet X elsewhere.

Use $Q(\gamma_5)$ to determine a double section $\tilde{Q}(\gamma_5)$ of $S(X)$. By the above, $\tilde{Q}(\gamma_5) \sim 2\vartheta + 2K_X + 2\kappa(\gamma_5)$ on $S(X)$, and is $N(\gamma_5)$-invariant. In fact, since the second projection $\pi_2 : \mathbb{P}(V) \times \mathbb{P}(V) \to \mathbb{P}(V)$ is of degree 2 when restricted to $S(X)$ (the dual curve of X has degree 2), and since π_2 on $S(X)$ ramifies exactly of order 2 along ϑ (X has no flex points), one sees that $\tilde{Q}(\gamma_5)$ must split into two local components at v_1 and $v_2 \in \vartheta \subset S(X)$. For all $z \in X$, $z \neq v_1$, v_2 , $\tilde{Q}(\gamma_5)$ intersects $S(X)_z$ in two distinct points (and therefore transversally); this is because the dual curves of X and $Q(\gamma_5)$ can only intersect with multiplicity 2 at the two tangents to v_1 and v_2. Hence, the desingularization of $\tilde{Q}(\gamma_5)$ is a 2 to 1 unbranched covering of X which is simply-connected. Therefore, $\tilde{Q}(\gamma_5)$ splits into two sections σ_1 , σ_2 and each $\sigma_i \sim \vartheta + K_X + \kappa(\gamma_5)$ on $S(X)$. Since $Q(\gamma_5)$ passes through the five fundamental points not fixed by $\langle\gamma_5\rangle$, $\tilde{Q}(\gamma_5)$ must pass through each $\infty_v \in S(X)_v$, for every $v \in \kappa$ not in $\kappa(\gamma_5)$. Since $\tilde{Q}(\gamma_5)\cdot\mathcal{V} = 24$, and we have found at least 24 points of intersection (counting multiplicities - recall that $\mathcal{V}$ has a cusp at each ∞_v) , we see that there are no other intersections, and that the local intersection number of $\tilde{Q}(\gamma_5)$ and $\mathcal{V}$ at ∞_v , $v \notin K(\gamma_5)$, is exactly 2 . We can conclude now that each component σ_i , i = 1, 2, of $\tilde{Q}(\gamma_5)$ intersects $\mathcal{V}$ transversally at $\kappa(\gamma_5)$, and with exact multiplicity 2 at each ∞_v in one of the two $\langle\gamma_5\rangle$ orbits in $\{\infty_v | v \notin \kappa(\gamma_5)\}$. We also conclude that any element of order 2 in $N(\gamma_5)$ interchanges σ_1 and σ_2 .

From what we know about local intersection multiplicities of the σ_i and $\mathcal{V}$, it follows the proper transforms $\hat{\sigma}_i$ in $\hat{S}(X)$ are disjoint from $\hat{\mathcal{V}}$ and $\hat{\vartheta}$. Hence, the branched covering $E(X) \to \hat{S}(X)$ splits over $\hat{\sigma}_i$. Performing this construction for each of the six $\langle\gamma_5\rangle$'s in G , we get twelve sections of $S(X)$, and by the splitting argument just given, we get 24 sections of $E(X)$, mutually disjoint, and disjoint from the 0-section. It now follows from the numerical formulas in [7], §1, that all sections of $E(X)$ are torsion points in the group structure over X .

Since the singular fibers are all of type I_5 , they must all be 5-torsion points, so that our 25 global sections are the group of all 5-torsion points, which splits over X .

Example 2: The Klein Quartic

The minimal non-trivial representations of $G = PSL(2,7)$ are three dimensional, say V and V^* , which are non-isomorphic. The minimal degree homogeneous form invariant by G on V is in $S^4(V^*)$, say F_4 , which is unique up to a constant factor. Let $X = \{F_4 = 0\}$ in $\mathbb{P}(V)$. The next degree invariant is $F_6 \in S^6(V^*)$, which is similarly unique: let Y denote its zero locus. It follows, as in example 1, by Hurwitz's theorem, that X and Y must be irreducible and non-singular. Furthermore, X has one orbit of fixed points of degree 2, one of degree 3 and one of degree 7, and $G\backslash X$ has genus 0 . This last orbit has minimal cardinality 24, and using the same idea as in example 1, we'll seek a G-invariant bihomogeneous form $P(z,\zeta)$ in $S^3(V^*) \otimes S^3(V^*)$. First note that

$$S^3(V^*) \simeq V^* \oplus V_7$$

as G-representation space. Here V_7 is the unique irreducible of dim 7 for G . It has a G-invariant quadratic form, and V does not, so there is a unique non-zero G-invariant , say $P(z,\zeta)$, in $S^3(V^*) \otimes S^3(V^*)$, up to a constant factor. (Unfortunately, I do not have a good geometric description as in example 1 for the family $P(z,\cdot)$ of cubics in $\mathbb{P}^2$.) Since $S^6(V^*)$ has a one-dimensional space of G-invariants also, and the multiplication map $S^3(V^*) \otimes S^3(V^*) \to S^6(V^*)$ is surjective, we conclude that $P(z,z)$ is not identically zero, and so $\{P(z,z) = 0\}$ must be the G-invariant sextic Y . Since $X \cdot Y = 24$, and the scheme of intersection is G-invariant, X and Y cross transversally at each point of $\kappa = \{\text{fixed points on } X \text{ of degree } 7\}$. Thus the triple section $\mathcal{V}$ of $S(X)$ determined by $P(z,\zeta)$ crosses $\theta \subset S(X)$ transversally exactly along κ . $\mathcal{V}$ is G-invariant and defining $0_v, \infty_v$ as above, for each $v \in \kappa$, we see that

(i) $\mathcal{V}$ is irreducible

(ii) $\mathcal{V}$ is non-singular except for a double point at each ∞_v , $v \in \kappa$, of the form $y^2 = x^5$.

(iii) for all $z \in X$, $z \notin \kappa$, ϑ intersects $S(X)_z$ in three distinct points.

This again follows from Hurwitz's formula and the genus formula, using G-invariance. Hence, to resolve the singularities of $\vartheta \cup \mathcal{O}$ in $S(X)$, we must blow-up each $v \in \kappa \subset \mathcal{O}$ and each ∞_v, $v \in \kappa$, twice. The new surface $\hat{S}(X)$ has singular fibers:

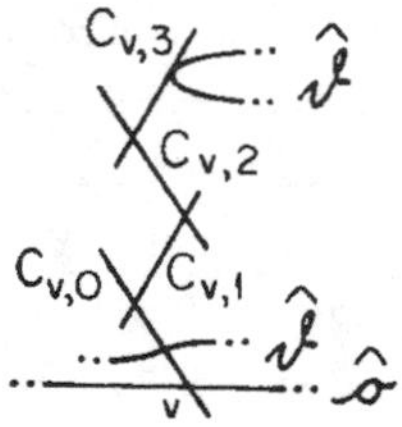

Since $\vartheta \sim 3\mathcal{O} + 3K_X + \kappa$ again, we see that

$$B = \hat{\mathcal{O}} + \hat{\vartheta} \sim 4\hat{\mathcal{O}} + 3K_X + \kappa - 2\cdot\sum_{v\in\kappa} C_{v,1} - 4\cdot\sum_{v\in\kappa} C_{v,2} - 6\cdot\sum_{v\in\kappa} C_{v,3}\,.$$

Hence to divide the divisor class B by two in a G-invariant way, it suffices to divide $K_X + \kappa$ on X. Since $\kappa \sim 6K_X$ on X, it suffices to divide K_X invariantly by 2, i.e., to find a G-invariant θ-characteristic. We first prove that such a G-invariant θ-characteristic, if it exists, is unique.

To see this, suppose $2\theta_i \sim K_X$, and that $\gamma_7^*(\theta_i) \sim \theta_i$, for $i = 1, 2$ and some $\gamma_7 \in G$ of order 7. We claim $\theta_1 \sim \theta_2$. To see this, note that otherwise the class of $\theta_1 - \theta_2$ is a $\langle\gamma_7\rangle$-invariant 2-torsion point in $\mathrm{Pic}(X)$. Note also that the normalizer $N(\gamma_7)$ of $\langle\gamma_7\rangle$ in G is of order 21, so that each γ_7 has three fixed points on X (all $\langle\gamma_7\rangle$'s in G are conjugate). But then Hurwitz's formula shows that $\langle\gamma_7\rangle\backslash X$ has genus zero, which implies $\mathrm{Pic}(X)$ has no non-zero $\langle\gamma_7\rangle$-invariant ℓ-torsion, for ℓ prime to 7.

To construct an invariant θ-characteristic seems quite difficult, and we give it explicitly here, but our proof of the G-invariance of its divisor class is indirect, by a counting argument. We first exhibit two types of θ-characteristic on X.

Recall that X, of genus 3, has 64 θ-characteristics. The 28 odd (or effective) θ-characteristics are in one-one correspondence with the 28 bitangents of X and the 28 dihedral subgroups of order 6, D_6, in G : each cyclic subgroup $\langle\gamma_3\rangle \subset G$ of order 3 has normalizer $N(\gamma_3) \simeq D_6$, and two fixed points P_1, P_2 on X. The line in $\mathbb{P}^2$ through those two points is tangent at P_1 and P_2 (otherwise D_6-invariance forces too high an intersection with the quartic X). Hence, $P_1 + P_2$ is a θ-characteristic with exact isotropy group $N(\gamma_3)$ for its divisor class.

Next, there are 24 fixed points of degree 7 on X. For an element $\gamma_7 \in G$ of order seven, $N(\gamma_7)$ has order 21. Fixing a $\langle\gamma_7\rangle$, there are three points on X fixed by $\langle\gamma_7\rangle$; call this divisor $\kappa(\gamma_7)$. $N(\gamma_7)$ permutes these three points cyclicly, and acts simply transitively on the other 21 points in κ. The triangle in $\mathbb{P}^2$ formed by the tangents to X at the points of $\kappa(\gamma_7)$ is invariant by $\langle\gamma_7\rangle$, hence its vertices must occur at the points of $\kappa(\gamma_7)$. Labelling the points $\kappa(\gamma_7)$ suitably, we get the picture:

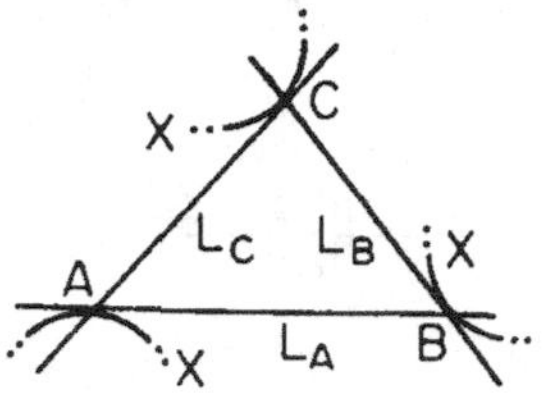

Since X is canonically embedded, we have

$$K_X \sim 3A + B \sim 3B + C \sim 3C + A$$

Hence,

$$K_X \sim 3A + B - (3B+C) + 3C + A = 4A - 2B + 2C .$$

Define $\theta_A = 2A - B + C$; we see $2\theta_A \sim K_X$. Similarly, $\theta_B = 2B - C + A$, and $\theta_C = 2C - A + B$ are also θ-characteristics. Each is clearly $\langle\gamma_7\rangle$-invariant. $N(\gamma_7)$ permutes θ_A, θ_B, θ_C. Note, finally, that

$$\begin{aligned}\theta_A - \theta_B &= 2A - B + C - (2B - C + A)\\ &= (3C + A) - (3B + C) \sim 0 .\end{aligned}$$

Hence, θ_A is a divisor class invariant by $N(\gamma_7)$. If A', B', C' are the vertices of the triangle associated to another $\langle\gamma_7'\rangle$, we get $\theta_{A'}$, invariant by $N(\gamma_7')$, and there are at most 8 distinct such θ_A's. By a previous remark, if there is a G-invariant θ-characteristic on X, it must be θ_A and all $\theta_{A'}$ are $\sim\theta_A$. Thus, either there are eight distinct θ_A's or just one which is G-invariant (because $N(\gamma_7)$ is a maximal proper subgroup of G). For the same reason, no θ_A is equivalent to one of the 28 odd θ-characteristics.

<u>Lemma</u>: θ_A is a G-invariant divisor class.

<u>Proof</u>: The proof is by contradiction. If θ_A is <u>not</u> G-invariant, by the above we have constructed 28 + 8 = 36 of the 64 θ-characteristics. Let us consider how it is possible to partition the remaining 28 into G-orbits. The possible isotropy subgroups H in G with $\#H\backslash G \leq 28$ are as follows:

H	$\#H\backslash G$
$N(\gamma_3) \simeq D_6$	28
$\langle\gamma_7\rangle$	24
$N(\gamma_2) \simeq D_8$	21
A_4	14
$N(\gamma_7)$	8
O_{24}	7

Here O_{24} denotes a group isomorphic to the group of rotations of the regular octohedron in $\mathbb{R}^3$; there are two conjugacy classes of such subgroups in G, interchanged by the unique outer automorphism of G. Thus, the G-orbits give a partition

$$28 = \alpha\cdot 7 + \beta\cdot 8 + \gamma\cdot 14 + \delta\cdot 21 + \varepsilon\cdot 24 + \zeta\cdot 28 , \qquad (4.1)$$

where all coefficients are non-negative integers. We want to show that (4.1) leads to a contradiction. Obviously, ε must be 0, and thus β must be 0, since it must be divisible by 7. Thus,

$$28 = \alpha\cdot 7 + \gamma\cdot 14 + \delta\cdot 21 + \zeta\cdot 28 . \qquad (4.2)$$

Note that, for any subgroup H of G, the set of θ-characteristics invariant by H, if non-empty, is in bijective correspondence with H-invariant 2-torsion in $\mathrm{Pic}(X)$. To use this, consider $\langle\gamma_3\rangle \subset G$, and $Y = \langle\gamma_3\rangle\backslash X$. Since $\langle\gamma_3\rangle$ has 2 fixed points of degree 3 on X, Y has genus 1. $\langle\gamma_3\rangle\backslash N(\gamma_3) \simeq \mathbb{Z}/2$ acts on Y, and the non-trivial involution has 4 fixed points on Y. This last is because each γ_2 in G has 4 fixed points in X. This gives the four $\langle\gamma_3\rangle\backslash N(\gamma_3)$-invariant two-torsion points in $\mathrm{Pic}(Y)$, and hence exactly four $N(\gamma_3)$-invariant 2-torsion points in $\mathrm{Pic}(X)$.

Since any D_6 in G is contained in exactly one O_{24} in each conjugacy class of such groups, and D_6 is its own normalizer in G, the preceeding calculations impose the condition

$$4 = 1 + \alpha + \zeta \qquad (4.3)$$

in addition to (4.2). These two equations can't be solved for non-negative, integral $\alpha, \gamma, \delta, \zeta$. This contradiction implies θ_A is G-invariant.

As before, we can calculate the numerical invariants of the resulting $E(X)$. It has an action of $\hat{G} = SL(2,7)$, and all sections must be 7-torsion sections. It appears difficult to describe these sections directly, geometrically. For example, the image of such a section in $\mathbb{P}^2$ will be a $N(\gamma_7)$-invariant curve of degree 11 passing through the points of κ in a special way.

Example 3: Bring's curve

Bring's curve is the $G = A_5$-invariant sextic in $\mathbb{P}^2$ passing through the six fundamental points of $\mathbb{P}^2$ for the G-action. The sextic has six nodes at the fundamental points. We let X denote the desingularization of the sextic, and $S(X)$ the normalization of its tangential ruled surface in $\mathbb{P}^2 \times \mathbb{P}^2$. As noted earlier, X is of genus 4 and there are 2 orbits of fixed points of degree 5, and one orbit of fixed points of degree 2. In example 1 above we also noted that X is the complete intersection of the invariant quadric and cubic in $\mathbb{P}(V_4)$, where V_4 is the unique irreducible G representation of dim 4. Consider $\mathbb{C}_5$ with the permutation representation of

$G = A_5$. In terms of the standard coordinates $x = (x_1, \ldots, x_5)$,
set $H_n(x) = \sum_{i=1}^{5} x_i^n$. Then V_4 is the invariant subspace given by
$\{H_1(x) = 0\}$. The homogeneous equations for the quadric and cubic
in $\mathbb{P}(V_4)$ are $\{H_1 = H_2 = 0\}$, and $\{H_1 = H_3 = 0\}$, respectively,
while X is defined by $\{H_1 = H_2 = H_3 = 0\}$. Note that all of the
above equations are invariant under the full permutation group $\mathfrak{S}_5$,
permuting the coordinates. Note also that $X \subset \mathbb{P}^3$ is canonically
embedded, and that the invariant quartic $\{H_4 = 0\}$ meets X precisely
along the union of the two orbits of fixed points of degree 5 on
X (24 points). Any transposition $\sigma \in \mathfrak{S}_5$ will interchange these two
A_5-orbits of degree 5 fixed points.

To construct an elliptic surface over X , consider $S(X)$ immersed
in $\mathbb{P}^2 \times \mathbb{P}^2$. Since A_5 has an invariant quadric Q in $\mathbb{P}^2$, one
constructs an A_5-invariant section $\mathcal{S}_1$ of $S(X)$ using this form.
Algebraically, one considers the bihomogeneous form of bi-degree
$(1,1)$ in $V_3^* \otimes V_3^*$ corresponding to the invariant bilinear form. More
geometrically, $\mathcal{S}_1$ crosses L_z , $z \in X$, at the point where L_z and
the Q-polar of z intersect (viewing X as immersed in $\mathbb{P}^2$).
$\mathcal{S}_1$ meets $\mathcal{O}$ transversally at the degree 5 fixed points not at the
fundamental points of the sextic. Call the set of these points κ_1 ,
the set of degree 5 fixed points at the fundamental points κ_2 , and
set $\kappa = \kappa_1 + \kappa_2$. To get an A_5-invariant double section, consider
the quadric Q itself as a bihomogeneous form of bidegree $(0,2)$,
in $S^0(V_3^*) \otimes S^2(V_3^*)$. This defines an invariant double section $\mathcal{S}'$
which crosses $\mathcal{O}$ again exactly at κ_1 . Motivated by §3 above, note
that any transposition $\sigma \in \mathfrak{S}_5$ acts on $S(X)$, and that $\mathcal{S}_2 =_{\mathrm{def}} \sigma(\mathcal{S}')$
meets $\mathcal{O}$ transversally at κ_2 , and is A_5-invariant. Finally, define
$\mathcal{S} = \mathcal{S}_1 + \mathcal{S}_2$.

Note, first, that $\mathcal{S}_2$ is irreducible. Otherwise, each component
would be an A_5-invariant section, and only one of them would meet $\mathcal{O}$
along the A_5-orbit κ_2 . Thus, the other component would be a section
of $S(X)$ disjoint from $\mathcal{O}$, contradicting the non-triviality of $S(X)$.
As earlier, using both the genus formula for $\mathcal{S}_2 \subset S(X)$, and Hurwitz's
formula for the double covering $\mathcal{S}_2 \to X$, one concludes that $\mathcal{S}_2$ has
only 12 cusp singularities at the points ∞_v , $v \in \kappa_1$, and that $\mathcal{S}_2$
intersects all other fibers of $S(X)$ in two distinct points.

To find the remaining singularities of $\mathcal{S}$, observe that $\mathcal{S}_1$
and $\mathcal{S}_2$ both pass through ∞_w , for any $w \in \kappa_2$. The local equation
for $\mathcal{S}$ near ∞_w must be, in suitable local coordinates, $y^2 = x^\nu +$
higher order terms, where $\nu \equiv 3 \bmod 5$. (This follows from γ_5-invar-

iance, for $\langle\gamma_5\rangle = G_w$.) Since ϑ splits at ∞_w, $\nu \geq 8$, and the local intersection multiplicity of ϑ_1 and ϑ_2 is ≥ 4 at ∞_w. Thus, we've calculated that, globally, $\vartheta_1 \cdot \vartheta_2 \geq 48$. However, $\vartheta_1 \sim \sigma + K_X + \kappa_1$, $\vartheta_2 \sim 2 + 2K_X + \kappa_2$, so $\vartheta_1 \cdot \vartheta_2 = 48$, and the local multiplicity of $\vartheta_1 \cdot \vartheta_2$ at ∞_w is exactly 4, and ϑ_1, ϑ_2 are disjoint except at these ∞_w points.

To resolve the singularities of $\vartheta + \sigma$ we must blow-up i) the points in $\kappa \subset X = \sigma \subset S(X)$; ii) the points ∞_v, $v \in \kappa_1$; and iii) the points ∞_w, $w \in \kappa_2$, four times. The resulting singular fibers of the blow-up $\hat{S}(X)$ over X look like:

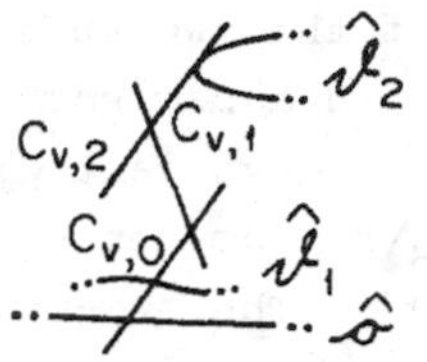

over $v \in \kappa_1$, and like:

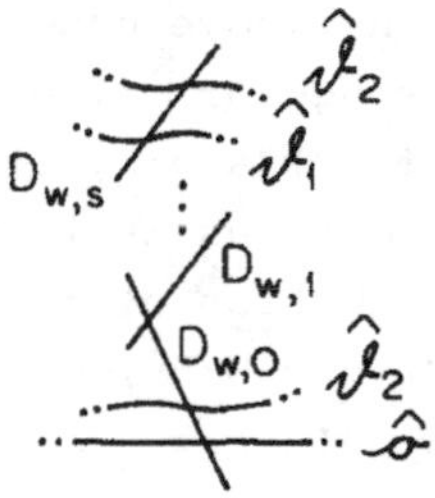

over $w \in \kappa_2$. (The negative integers are self-intersection numbers; the $C_{v,i}$, $D_{w,i}$ label the components). In terms of these cycles one finds easily that

$$\hat{\sigma} + \hat{\vartheta} \sim 4\hat{\sigma} + 3K_X + \kappa - \sum_{\substack{v\in\kappa_1 \\ i=1,2}} 2i\, C_{v,i} - \sum_{\substack{w\in\kappa_2 \\ j=1,\dots,5}} 2j\, D_{w,j}\,.$$

Since we saw above that $\kappa = \kappa_1 + \kappa_2 \sim 4K_X$, to find an A_5-invariant divisor class D with $2D \sim \hat{\vartheta}^1 + \vartheta^2$, it suffices to find an A_5-invariant θ-characteristic. By Hurwitz's theorem, the quotient $\langle\gamma_5\rangle\backslash X$ is rational, so there exists at most one such θ-characteristic.

To construct a θ-characteristic consider X as a curve in $\mathbb{P}^3$ again, and then as a curve on the quadric $\tilde{Q} = \{H_2 = 0\}$. X had bidegree $(3,3)$ on $\tilde{Q}$, and the two rulings of $\tilde{Q}$ give two A_5-equivariant maps $X \to \mathbb{P}^1$ of degree 3. Let L and L' denote the corresponding A_5-invariant divisor classes of degree 3 on X. Note that L and L' are interchanged by a transposition $\sigma \in G_5$. Next note that $L + L' \sim K_X$: at a point $q \in X$, consider the projective tangent plane to $\tilde{Q}$ at q, $T_q(\tilde{Q})$. $T_q(\tilde{Q})$ intersects $\tilde{Q}$ in two lines ℓ, ℓ' meeting at q, one from each of the two rulings. In $\tilde{Q}$, ℓ meets X in a divisor L, ℓ' meets X in a divisor L'; thus, $L + L' \sim K_X$. Finally, note that $L - L'$ is an A_5-invariant divisor of degree 0, hence is a torsion class, since $A_5\backslash X$ has genus zero. In fact, since $\langle\gamma_5\rangle\backslash X$ is rational, $5(L - L') \sim 0$. The A_5-invariant θ-characteristic on X is therefore given explicitly as $D = L + 3(L' - L)$: it is clearly A_5-invariant, and $2D = 2L + 6(L' - L) \sim 2L + L' - L \sim K_X$.

For this example, as for example 2, it appears that calculating explicitly the sections of the above constructed elliptic surface $E(X)$ is too involved because of the high degrees of the invariant forms required. We simply note that, by the above construction, $E(X)$ has I_5-singular fibers over κ_1, I_{10}-singular fibers over κ_2, and no others. Hence. using the results of [7] again, all sections of X are torsion sections, and the group of sections is isomorphic to a subgroup of $\mathbb{Z}/10 \oplus \mathbb{Z}/5$. We already have a section of order **2** (viz., $\hat{\vartheta}_1$). The group of sections is exactly $\mathbb{Z}/10 \oplus \mathbb{Z}/5$, which can be seen in terms of cohomology calculations on $S(X)$.

References

[1] R. Fricke, Lehrbuch der Algebra, Bd. II, Braunschweig, 1926.

[2] E. Hecke, Die eindeutige Bestimmung der Modulfunktionen q-ter Stufe durch algebraische Eigenschaften, Math. Ann. 111 (1935), 293 - 301.

[3] F. Klein, Weitere untersuchungen über das Ikosaeder, Gesammelte Math. Abh. II, 321 - 384.

[4] K. Kodaira, On compact analytic surfaces, II: Ann. Math. 77 (1963), 563 - 626; III: ibid. 78 (1963), 1 - 40.

[5] G.A. Miller, H.F. Blichfeldt, L.E. Dickson, Theory and Application of Finite Groups, New York, 1936.

[6] I. Naruki, Über die Kleinsche Ikosaeder - Kurve sechsten Grades, Math. Ann. 231 (1978), 205 - 216.

[7] T. Shioda, On elliptic modular surfaces, J. Math. Soc. Japan 24 (1972), 20 - 59.

ON THE RATIONALITY OF CERTAIN MODULI SPACES RELATED TO CURVES OF GENUS 4

F. Catanese* - Dip. di Matematica
Università di PISA, via Buonarroti 2

INTRODUCTION

Let M_g be the coarse moduli space for complete smooth curves of genus g, let R_g be the "Prym moduli space" of unramified (connected) double covers of curves of genus g; a general problem is: what can be said about the birational structure of M_g, R_g?

From the point of view of birational geometry we can also talk about $M_{g,n}$, the moduli space of curves of genus g together with an ordered n-tuple of points, though this moduli functor is not representable in general (cf. [10]).

Our main results are:

<u>Theorem A</u>: R_4 is a rational variety.

<u>Theorem B</u>: $M_{4,1}$ is a rational variety.

<u>Theorem C</u>: M_4 admits a covering of degree 24 by a rational variety.

To put these results into perspective, we notice that, while the rationality of M_1, R_1 is classical and well-known, the rationality of M_2 has been proved by Igusa (cf. [8], also [17]).

For higher values of the genus g, the situation is as follows:

i) M_g is known to be unirational for $g \leq 10$, ([16], [1]), $g = 12$ ([14]), uniruled for $g = 11$ ([9]), whereas, for g odd ≥ 25 M_g is variety of general type ([7]), and D. Mumford and J. Harris announced a similar result also for g even ≥ 40

ii) the unirationality of R_g for $g = 5,6$ has been proven only recently ([4], [6]).

If the base field is of characteristic $\neq 2$, R_g is a covering of M_g of degree $2^{2g} - 1$, so that theorems A and C produce two rational coverings of

---

*) Part of this research was done when the author was at the Institute for Advanced Study, partially supported by NSF grant MCS 81-033 65.
The author is a member of G.N.S.A.G.A. of C.N.R.

M_4 of degrees, respectively, 255 and 24.

I should finally remark that it is conjectured that M_3, M_4 are rational varieties, but, to my knowledge, this is still unsolved.

Our method of proof does not use classical invariant theory: our strategy consists in constructing, using the geometry of curves of genus 4, some rational Galois covers by some rational variety, and then computing explicitly the subfield of invariants.
These covers are constructed with elementary arguments in the case of theorems B),C) For theorem A) I use a classical result of Wirtinger ([18]), which can also be found in [5], and about which I was told by S. Recillas, (cf. [13]), to whom I am indebted for noticing a mistake in an earlier proof of Theorem 1.5.

Our notation is as follows:
k is an algebraically closed field of char. $\neq 2$
X is a complete smooth curve of genus g defined over k
Pic(X) is the group of divisors on X modulo linear equivalence, here denoted by $\equiv$.
η is a divisor in $Pic_2(X) - \{0\}$, i.e. $2\eta \equiv 0$, $\eta \not\equiv 0$.
K is a canonical divisor on any Gorenstein variety Y, i.e. $\mathcal{O}_Y(K_Y) \cong \omega_Y$.
If D is a divisor, $|D|$ is the linear system of effective divisors $D' \equiv D$.
$\underline{S}_n$ is the symmetric group in n letters, and
V_n is the standard (permutation) representation on k^n.
Given any coherent sheaf F on a complete variety Y, we denote by $h^i(F)$ the dimension of $H^i(Y,F)$ as a k-vector space.
If U is a k-vector space, we denote by U^* its dual space.
R.R. is an abbreviation for the Riemann-Roch theorem

§1. GEOMETRY OF CURVES OF GENUS 4.

Let X be a non-hyperelliptic curve of genus 4. Then the linear system $|K_X|$ gives an embedding of X in $\mathbb{P}^3$ such that the image of X is the complete intersection of a quadric Q and of a cubic G.

The quadric Q is uniquely determined, and, since it is normal, as well as G, there are only two possibilities:

i) Q is smooth

ii) Q is a quadric cone.

Case ii) occurs if and only if there exists a half canonical divisor ν (i.e. $2\nu \equiv K_X$) such that $h^0(\mathcal{O}_X(\nu)) = 2$: we say then that X has a vanishing thetanull.

It is well-known that M_4 is an irreducible variety of dimension 9 and that curves with a vanishing thetanull form an 8-dimensional subvariety, hyperelliptic curves form a 7-dimensional subvariety.

Definition 1.1. Let $\eta \in \mathrm{Pic}_2(X) - \{0\}$. We shall say that the pair (X,η) is bi-elliptic if there exist an elliptic curve E, a double covering $f:X \to E$, and a divisor $\eta' \in \mathrm{Pic}_2(E) - \{0\}$ such that $\eta \equiv f^*(\eta')$.

Definition 1.2. A normal cubic surface G in $\mathbb{P}^3$ is said to be symmetric if its equation can be written as the determinant of a symmetric 3×3 matrix of linear forms (cf. [2]). A symmetrization of G is the datum of such a matrix $(a_{ij}(y)) = (a)$, where $y = (y_0,y_1,y_2,y_3)$ are coordinates in $\mathbb{P}^3$, up to the action of $PGL(3)$ (such that, for $g \in GL(3)$, $(a) \longmapsto {}^tg(a)g$).

How many symmetric cubics with a symmetrization are there in $\mathbb{P}^3$, up to the action of $PGL(4)$?

The answer is: as many as there are pencils of conics in $\mathbb{P}^2$, up to the action of $PGL(3)$.

In fact, let U be the space $\mathrm{Sym}^2(k^3)$ of symmetric 3×3 matrices; then $\mathbb{P}(U)$ is the space of conics in $\mathbb{P}^2$, and $\mathbb{P}(U)$ contains the cubic determinantal hypersurface $\Delta = \{\det(a_{ij}) = 0\}$: Δ is the dual variety of the Veronese surface W^* in $\mathbb{P}(U^*)$, and its singular locus is the Veronese surface W in $\mathbb{P}(U)$.

Now, the datum of a symmetrization amounts to giving a $\mathbb{P}^3 \subset \mathbb{P}(U)$ such that $\mathbb{P}^3 \cap \Delta$ is a normal cubic. But giving a $\mathbb{P}^3 \subset \mathbb{P}(U)$ is equivalent to giving a $\mathbb{P}^1$ in $\mathbb{P}(U^*)$, i.e. a pencil of conics.

Notice that the number of base points in the pencil of conics is the cardinality of $\mathbb{P}^3 \cap W$, the number of degenerate conics in the pencil is the cardinality of $\mathbb{P}^1 \cap \Delta^*$.

The following is the list of pencils of conics (up to projective equivalence):

i) pencils of reducible conics: $\lambda x_1^2 + \mu x_1 x_2 = 0$, or $\lambda x_1 x_2 + \mu x_2 x_3 = 0$

ii) pencil with 4 base points: $\lambda x_1 x_2 + \mu x_3 (x_1 + x_2 + x_3) = 0$

iii) pencil with 3 base points: $\lambda x_1 x_2 + \mu x_3 (x_1 - x_2) = 0$

iv) pencil with 2 base points, 2 degenerate conics: $\lambda x_1 x_2 + \mu x_3^2 = 0$

v) pencil with 2 base points, one reducible conic: $\lambda x_1 x_2 + \mu (x_1 x_3 - x_2^2) = 0$

vi) pencil with 1 base point: $\lambda x_1^2 + \mu (x_1 x_3 - x_2^2) = 0$.

Correspondingly we get the following symmetrizations:

i) $\begin{pmatrix} 0 & 0 & y_1 \\ 0 & y_0 & y_2 \\ y_1 & y_2 & y_3 \end{pmatrix}, \begin{pmatrix} y_0 & 0 & y_3 \\ 0 & y_1 & 0 \\ y_3 & 0 & y_2 \end{pmatrix}$ $\quad G \equiv \{y_0 y_1^2 = 0\}$, respectively

$G \equiv \{y_0 y_1 y_2 - y_1 y_3^2 = 0\}$, so G is reducible, and this case must be excluded,

ii) $\begin{pmatrix} y_0 & 0 & y_2 \\ 0 & y_1 & y_3 \\ y_2 & y_3 & (-y_2 - y_3) \end{pmatrix}$ $\quad G \equiv \{y_0 y_1 (y_2 + y_3) + y_0 y_3^2 + y_1 y_2^2 = 0\}$

Here G has 4 singular points, and is also projectively equivalent to the 4-nodal cubic of Cayley of equation $\sigma_3(y) = \sum_{i=0}^{3} \frac{y_0 y_1 y_2 y_3}{y_i} = 0$.

iii) $\begin{pmatrix} y_0 & 0 & y_2 \\ 0 & y_1 & y_2 \\ y_2 & y_2 & y_3 \end{pmatrix}$ $\quad G \equiv \{y_0 y_1 y_3 - y_0 y_2^2 - y_1 y_2^2 = 0\}$.

G has three singular points, two nodes and a singularity of type A_3 at $\{y_0 = y_1 = y_2 = 0\}$.

iv) $\begin{pmatrix} y_0 & 0 & y_2 \\ 0 & y_1 & y_3 \\ y_2 & y_3 & 0 \end{pmatrix}$ $\quad G \equiv \{y_0 y_3^2 + y_1 y_2^2 = 0\}$.

The line $y_3 = y_2 = 0$ is singular, so this case must be excluded.

v) $\begin{pmatrix} y_0 & 0 & y_1 \\ 0 & y_1 & y_2 \\ y_1 & y_2 & y_3 \end{pmatrix}$ $\qquad G \equiv \{y_0y_1y_3 - y_0y_2^2 - y_1^3 = 0\}.$

G has two singular points, one node at $\{y_1 = y_2 = y_3 = 0\}$, and a singular point of type A_5 at $\{y_0 = y_1 = y_2 = 0\}$.

vi) $\begin{pmatrix} 0 & y_0 & y_1 \\ y_0 & y_1 & y_2 \\ y_1 & y_2 & y_3 \end{pmatrix}$ $\qquad G \equiv \{y_3y_0^2 + y_1^3 - 2y_0y_1y_2\}$

The line $y_0 = y_1 = 0$ is singular: moreover the plane $\{y_0 = 0\}$ is in the tangent cone at every point of the singular line so that this cubic is not projectively equivalent to the one in iv).

Theorem 1.3. Every symmetric cubic G has only one symmetrization. Moreover, G has exactly one finite irreducible double cover ramified exactly at the singular points of G.

Proof. The first statement follows from the above list. Now, let $Y \xrightarrow{f'} G$ be a double cover, and notice that G has only singularities of type A_n, $n = 1,3,5$. The fibre product $Z' = Y \times_G \tilde{G}$, where $\tilde{G}$ is a minimal desingularization of G, is an irreducible finite cover of $\tilde{G}$.
Therefore, if Z is the normalization of Z', there exists a reduced effective divisor E with support in the exceptional divisor of $p:\tilde{G} \to G$, and a divisor L on $\tilde{G}$ such that $2L \equiv E$, and $f:Z \to \tilde{G}$ is the double cover of $\tilde{G}$ in $\mathcal{O}_{\tilde{G}}(L)$ branched over E. Hence Z has only nodes as singularities, and $\omega_Z = f^*(-H) + f^*(L)$, where H is an hyperplane section of $\tilde{G}$. Z is then a rational variety, therefore $f'_*\mathcal{O}_Y$ is a Cohen-Macaulay sheaf on $\mathbb{P}^3$, with support on G (cf. [2], prop. 2.18). It follows also, by the Riemann-Roch theorem, that $h^0(\mathcal{O}_{\tilde{G}}(H-L)) = 3$. Applying theorem 2.19 of [2] we prove that a double cover as above gives a symmetrization.
Conversely, consider the sheaf F cokernel of $0 \to \mathcal{O}^3_{\mathbb{P}^3} \xrightarrow{(a_{ij}(y))} \mathcal{O}_{\mathbb{P}^3}(1)^3 \to F \to 0$, and define Y to be $\mathrm{Spec}(\mathcal{O}_G \oplus F)$, with algebra structure given as in [2], cor. 2.17.

Q.E.D.

We observe now that if we write $X = Q \cap G$, where G is symmetric, we are giving, as X is smooth, an unramified double cover $\tilde{X}$ of X, induced from the double cover $Y \to G$, or, equivalently a divisor $\eta \in \mathrm{Pic}_2(X)$. Now η is not trivial if $h^0(X, \mathcal{O}_X(K+\eta)) = 2$, and this follows from the exact sequence

$$(1.4) \qquad 0 \to H^0(\mathcal{O}_{\tilde{G}}(-H+L)) \to H^0(\mathcal{O}_{\tilde{G}}(H+L)) \to H^0(\mathcal{O}_X(K+\eta)) \to 0.$$

We are going now to prove a converse to this statement,

Theorem 1.5. (Wirtinger-Coble-Recillas).

Let X be a curve of genus 4, not hyperelliptic and with no vanishing theta null. Then giving a divisor $\eta \in \mathrm{Pic}_2(X) - \{0\}$, such that (X,η) is not bielliptic, is equivalent to writing X as the complete intersection of a smooth quadric Q and a symmetric cubic G.

Proof. We have already proven that giving G symmetric containing X determines an $\eta \in \mathrm{Pic}_2(X) - \{0\}$.
Conversely, consider the rational mapping $\Phi: X \to \mathbb{P}^2$ given by the linear system $|K_X + \eta|$. We break up the proof in several steps.

(1.6) $|K + \eta|$ has no base points if X is not hyperelliptic.

Proof. Let $p \in X$: since $H^1(\mathcal{O}_X(K+\eta)) = 0$ p is not a base point if and only if $H^1(\mathcal{O}_X(K+\eta-p)) = 0$. By Roch's duality, this is equivalent to $|p-\eta| \neq \emptyset$. But if $q \in |p-\eta|$, then $2p \equiv 2q$, with $q \neq p$, and X is hyperelliptic.

Q.E.D.

Then Φ is a morphism. Denote by $C = \Phi(X)$, so that $\deg C \cdot \deg \Phi = 6$

(1.7) $\deg C \geq 3$ if X is not hyperelliptic.

Proof. If C is a smooth conic, then, let D be the inverse image of a general point in C: we have $h^0(\mathcal{O}_X(D)) \geq 2$, and D has degree 3. Let $D' = K - D$: by R.R. $h^0(\mathcal{O}_X(D')) = 2$, and $D' \not\equiv D$, since $2D \equiv K + \eta$, $\eta \not\equiv 0$. Since $|D|$ has no base points, by the "base point free pencil trick" (cf. [11]), $H^0(\mathcal{O}_X(K)) \cong$ $\cong H^0(\mathcal{O}_X(D')) \otimes H^0(\mathcal{O}_X(D))$. Since X is not hyperelliptic $H^0(\mathcal{O}_X(K))^{\otimes 2} \to H^0(\mathcal{O}_X(2K))$ is surjective. Since $2D' \equiv 2D$, it follows that then $H^0(\mathcal{O}_X(2D))^{\otimes 2} \to H^0(\mathcal{O}_X(2K))$ is surjective: this is anyhow absurd since $|2K|$ gives a birational morphism.

Q.E.D.

(1.8) If C is a singular cubic, then its normalization $\tilde{C}$ is $\mathbb{P}^1$, and, since Φ factors through $\tilde{C}$, X is hyperelliptic.

(1.9) C is a smooth cubic if and only if (X,η) is bielliptic.

Proof. If C is a smooth cubic, then $\Phi_* \mathcal{O}_X = \mathcal{O}_C \oplus \mathcal{O}_C(-E)$, $K_X = \Phi^*(E)$, where E is an effective divisor on C of degree 3. Let H be a hyperplane divisor on C. Then $\eta \equiv \Phi^*(H-E)$, but $\Phi^*:\mathrm{Pic}(C) \to \mathrm{Pic}(X)$ is injective ([12], pag. 332), hence (X,η) is bielliptic, with $\eta' \equiv H-E$.
Conversely, $K_X + \eta \equiv f^*(E+\eta')$, where $f_* \mathcal{O}_X \simeq \mathcal{O}_C \oplus \mathcal{O}_C(-E)$. But, by the Leray spectral sequence for the map f, $H^0(X,\mathcal{O}_X(K_X+\eta)) \cong f^*(H^0(C,\mathcal{O}_C(E+\eta')))$, therefore Φ factors through f and an embedding of C as a plane cubic.

Q.E.D.

Remark 1.10. An easy computation shows that bielliptic pairs form a six-dimensional subvariety of $\mathcal{R}_4$.

We are at the last step of the proof: $\Phi:X \to C$ is a birational morphism, therefore can be factored through a finite sequence of blow-ups.
We are therefore in the following situation: we are given a surface S obtained from $\mathbb{P}^2$ $(\Phi:S \to \mathbb{P}^2)$ by a finite sequence of blow ups of the (possibly infinitely near) singular points of C, of multiplicities $r_1 \geq r_2 \geq \dots r_k$.
Let $E_1, \dots E_k$ be the total transforms of the exceptional curves of each blow up, H the total transform of a line in $\mathbb{P}^2$.
Then, on S, we have

(1.11) $H^2 = 1, \quad H\cdot E_i = 0, \quad E_i^2 = -1, \quad E_i\cdot E_j = 0$ for $i \neq j$,

(1.12) $K_S \equiv -3H + \sum_{i=1}^{k} E_i, \quad X \equiv 6H - \sum_{i=1}^{k} r_i E_i.$

(1.13) Let Δ be $\sum_{i=1}^{k} (r_i-1)E_i$; by the adjunction formula $\mathcal{O}_X(K_X) = \mathcal{O}_X(3H-\Delta)$, and, on X, $\Delta \equiv 3H - K_X \equiv 2H + \eta$.

Therefore i) $H^0(\mathcal{O}_X(2H-\Delta)) = 0$, hence $H^0(\mathcal{O}_S(2H-\Delta)) = 0$

ii) $\Delta\cdot K_X = 12 = \sum_{i=1}^{k} r_i(r_i-1)$.

Since $H^1(S,K_S) = H^1(S,\mathcal{O}_S) = 0$, we have an isomorphism of $H^0(\mathcal{O}_S(3H-\Delta)) \to H^0(\mathcal{O}_X(K_X))$ given by restriction.

Let H' be the inverse image of a line not passing through the singular points of C. Since $H'(3H-\Delta) = 3$, the exact sequence

$$(1.14)\qquad 0 = H^o(\mathcal{O}_S(2H-\Delta)) \to H^o(\mathcal{O}_S(3H-\Delta)) \to H^o(\mathcal{O}_{H'}(3H-\Delta))$$

says that the rational map $\psi: S \to \mathbb{P}^3$ given by the linear system $|3H-\Delta|$ embeds H' as a twisted cubic.
Therefore, if $G = \psi(\mathbb{P}^2)$, G contains a 2-parameter family of twisted cubics, two of which intersect in only one point, so that G is not a smooth quadric.
If X has no vanishing thetanull, G must be a cubic surface.

Let F be the fixed part of the linear system $|3H-\Delta|$, so that $|3H-\Delta| = F + |M|$, $M^2 = \deg(G)$, $M\cdot F \ge 0$. Since ψ embeds a general line in $\mathbb{P}^2$, F is a sum of exceptional curves, each of which can be written either in the form E_i, or in the form $E_i - E_j$, $i > j$.
Then $M^2 = (F+M)^2 - (F+M)\cdot F - FM \le (3H-\Delta)^2 - (3H-\Delta)\cdot F$.
Now $(3H-\Delta)\cdot F = -\Delta F = -\sum_{i=1}^{k}(r_i-1)E_i\cdot F$, and for each component of F, $-\Delta\cdot F = 0$ or 1 $(-\Delta\cdot E_i = r_i - 1 \ge 1$, $-\Delta(E_i - E_j) = r_i - r_j \ge 0$ since $i > j)$.

Hence $M^2 \ge (3H-\Delta)^2 = 9 - \sum_{i=1}^{k}(r_i-1)^2 = 9 - \sum_{i=1}^{k} r_i(r_i-1) - \sum_{i=1}^{k}(r_i-1) = -3 + \sum_{i=1}^{k}(r_i-1)$.

If $\deg(G) = M^2 = 3$, then $r_1 = 2$; if $\deg(G) = 2$, then the only other possibility is $r_1 = 3$, $r_2 = 2$.

We can assume from now on $r_1 = 2$. By the exact sequence

$$(1.15)\qquad 0 \to \mathcal{O}_S(4H-2\Delta-X) \to \mathcal{O}_S(4H-2\Delta) \to \mathcal{O}_X \to 0$$

since $H^1(\mathcal{O}_S(4H-2\Delta-X)) = H^1(\mathcal{O}_S(-2H)) = 0$, we conclude that

(1.16) $|4H - 2\Delta|$ has dimension 0.

Let D be the unique divisor in $|4H-2\Delta|$: since $D \cap X = \emptyset$, D is mapped to the singular points of G, and, since $D \equiv 2(2H-\Delta)$, $|2H-\Delta| = \emptyset$, S admits a double covering ramified exactly on D; hence G admits a double covering ramified at most on the singular points, so that G is a symmetric cubic.
Since $\mathcal{O}_X(2H-\Delta) = \mathcal{O}_X(\eta)$, we have proven that G induces on X the double cover associated to η.
Conversely, given X as $Q \cap G$, let η be the induced divisor: then Φ is induced by the linear system $|H-L|$ on $\tilde{G}$. If H'' is a general hyperplane section of G, one has a restriction isomorphism of $H^o(\mathcal{O}_{\tilde{G}}(H-L)) \to H^o(\mathcal{O}_{H''}(H-L))$,

therefore, if we denote still by Φ the rational map $\Phi:\tilde{G} \to \mathbb{P}^2$ given by $|H - L|$, Φ embeds H'' as a smooth plane cubic. Since through any two general points x,y of G there passes a plane section H'' as above, Φ is birational, and the inverse map $\psi:\mathbb{P}^2 \to G$ is given by a system of plane cubics. Since $\Phi_{|X}$ is a morphism, clearly $|K + \eta|$ gives a birational morphism and (X,η), by (1.9), is not bi-elliptic.

Q.E.D.

Just for completeness, we indicate, for the three types of symmetric cubics, which are the systems of plane cubics giving the rational map ψ.
In case ii) we consider the six points of intersection of four independent lines in $\mathbb{P}^2$, and we blow then up to get $S \cong \tilde{G}$, with $\Delta = \sum_{i=1}^{6} E_i$, and $D \in |4H - 2\Delta|$ given by the of the proper transforms of the four lines (cf. e.g. [3]).
In case iii): take three lines L_1,L_2,L_3 in general position in $\mathbb{P}^2$ and blow up $\mathbb{P}^2$ at the three points $L_i \cap L_j$, at a fourth point $P_4 \in L_3$, and then at the 2 infinitely near points P_{4+i} lying over $L_i \cap L_3 = P_i$ $(i=1,2)$ in the direction of L_i. Let $P_3 = L_1 \cap L_2$.
Here you obtain S where $D \in |4H - 2\Delta|$ is given by the proper transform of $2L_3 + L_1 + L_2$ together with $E_1 - E_5$, $E_2 - E_6$, and $S \cong \tilde{G}$.
The double cover Z of S is smooth, being branched on the proper transforms of L_1, L_2, and (E_1-E_5) (E_2-E_6), i.e. on a smooth divisor consisting of four (-2) rational curves, while the finite cover Y has just a node as singularity, lying over the A_3 singular point of G.

Since we believe that case v) is the least known, we explain how to obtain the mapping ψ.

Choose w_0, w_1, w_2 a basis of $H^0(\mathcal{O}_{\tilde{G}}(H-L))$ such that (cf. [2], cor. 2.17) the following relations hold:

$$(1.17)\qquad \begin{cases} y_0w_0 + y_1w_2 = 0 \\ y_1w_1 + y_2w_2 = 0 \\ y_1w_0 + y_2w_1 + y_3w_2 = 0 \end{cases}$$

We can solve these as linear equations in $y_0, \ldots y_3$ and express then as homogeneous polynomials in (w_0,w_1,w_2).
We get $y_0 = w_2^3$, $y_1 = -w_0w_2^2$, $y_2 = w_0w_1w_2$, $y_3 = w_0(w_0w_2 - w_1^2)$, and this is an expression of ψ in appropriate coordinates on $\mathbb{P}^2$ and $\mathbb{P}^3$.

The system of cubics has 2 base points, namely $\{w_2 = w_0 = 0\} = P$, and $\{w_2 = w_1 = 0\} = P'$, and a general cubic of the system is smooth at P, P':

but to obtain a system free of base points one has to blow up three times over P at the points where the line $\{w_0 = 0\}$ (whose proper transform will be denoted by L_0) passes and three times over P' at the points where the conic $\{w_0 w_2 - w_1^2 = 0\}$ passes through.
Denote by L_2 the proper transform of the line $\{w_2 = 0\}$.
We get thus $E_1, E_2, E_3,$ E_1', E_2', E_3' on S, and we notice that $L_2^2 = L_0^2 = -2$, L_2 intersects transversally in exactly one point (E_1-E_2), $(E_2'-E_3')$; L_0 intersects E_3 transversally in exactly one point.
The total transform of the quartic $\{w_0 w_2^3 = 0\}$ is thus
$L_0 + (E_1+E_2+E_3) + 3(E_1'+E_2'+E_1) + 3L_2$ i.e. $3L_2 + L_0 + 2\Delta + (E_1'-E_2') + 2(E_2'-E_3')$ $+ 2(E_1-E_2) + (E_2-E_3)$.
The normal double cover Z of $S = \tilde{G}$ is thus ramified on
$L_2 + L_0 + (E_2-E_3) + (E_1'-E_2')$, hence Z is smooth, and the finite cover Y of G has just a singular point of type A_2 lying over the singular point of G of type A_5.

The meaning of theorem 1.5 in terms of $\mathcal{R}_4$ is the following

Theorem 1.18. $\mathcal{R}_4$ is an irreducible variety, birational to the quotient $\mathbb{P}(\mathrm{Sym}^2(V_4))/\underline{S}_4$, where V_4 is the standard representation of

Proof. Since $\mathcal{R}_4$ is a finite cover of $\mathcal{M}_4$, it is pure dimensional.
Let A be the open set of $\mathcal{R}_4$ corresponding to pairs (X,η) such that:

i) X is not hyperelliptic
ii) X has no vanishing thetanull
iii) (X,η) is not bielliptic.

By remark 1.10 and the considerations made at the beginning of the paragraph A is dense.

Let Q be a fixed smooth quadric in $\mathbb{P}^3$, and let B be the open set in the space of symmetric 3×3 matrices of linear forms such that, if $(a_{ij}(y)) \in B$, $G = \det(a_{ij}(y))$ is a normal cubic and $X = G \cap Q$ is a smooth curve of degree 6.

In view of theorem 1.5, there is a morphism of B onto A which is a quotient by the previously described action of $GL(3)$ on B. Hence $\mathcal{R}_4$ is irreducible (actually this was known already).

Moreover, let B' be the open subset of B such that G is a 4-nodal cubic (case ii)), and A' its image in $\mathcal{R}_4$: A' is again dense, being non-empty.

Assume that (X,η) corresponds to giving generators Q, G of the ideal of X in $\mathbb{P}^3$ such that G is a symmetric cubic, and analogously (X',η') corresponds to (Q',G'); if $f: X \to X'$ is an isomorphism such that $f^*(\eta') = \eta$, then

f is induced by a projectivity $g: \mathbb{P}^3 \to \mathbb{P}^3$ such that $Q = g^*(Q')$, $G = g^*(G')$, by theorem 1.5, and, conversely, such a projectivity induces an isomorphism of the pair (X,η) with the pair (X',η').
Since all 4-nodal cubics are projectively equivalent, we can fix the 4-nodal cubic to be G_0, the cubic of equation $\sigma_3(y) = \sum_{i=1}^{4} \frac{y_1y_2y_3y_4}{y_i} = 0$.

Consider now the open set Λ in $\mathbb{P}(\mathrm{Sym}^2(V_4))$ corresponding to the quadrics Q in $\mathbb{P}(V_4)) = \mathbb{P}^3$ such that $Q \cap G_0$ is a smooth sextic curve X. We get thus a morphism f of Λ into R_4, with $f(\Lambda) \supset A'$, such that Q, Q' map to the same pair (X,η) if and only if there exists $g \in PGL(4)$ such that $g(G_0) = G_0$, $g(Q) = Q'$. We conclude the proof since it is well-known that $\underline{S}_4$ is the group of projective automorphisms of G_0.

Q.E.D.

We want to find out now a dominant rational map of $\mathbb{P}^{10}$ to $M_{4,1}$.
To do this, recall that a curve of genus 4 X which is not hyperelliptic and has no vanishing thetanull is a smooth divisor of bidegree $(3,3)$ on $Q = \mathbb{P}^1 \times \mathbb{P}^1$.

Fix three points $\infty, 0, 1$ in $\mathbb{P}^1$ and let $p \in Q$ be the point (∞,∞), M be the (unordered) set of five points $\{(\infty,\infty), (\infty,0), (\infty,1), (0,\infty) (1,\infty)\}$.

Given a general $[C',p'] \in M_{4,1}$ we can assume to have chosen coordinates in $\mathbb{P}^1 \times \mathbb{P}^1$ such that $p = (\infty,\infty)$, and that the two lines $\{\infty\} \times \mathbb{P}^1$, $\mathbb{P}^1 \times \{\infty\}$ intersect C' in three distinct points. Let I_M be the ideal sheaf of M on Q. Therefore if we take the linear system $|G| = |I_M(3,3)|$ we obtain a rational dominant map of $|G|$ onto $M_{4,1}$ just by sending $C \in |G|$ to the pair $[C,(\infty,\infty)]$.

Assume now that two pairs $C, C' \in |G|$ are isomorphic: then there exists an automorphism g of $\mathbb{P}^1 \times \mathbb{P}^1$ which leaves (∞,∞) fixed and such that $g(C) = C'$, since all the automorphisms of $\mathbb{P}^1 \times \mathbb{P}^1$ are induced, via the Segre embedding, by automorphisms of $\mathbb{P}^3$.
But now g leaves the set $A = (\mathbb{P}^1 \times \{\infty\}) \cup (\{\infty\} \times \mathbb{P}^1)$ invariant, and, since $M = A \cap C = A \cap C'$, $g(M) = M$.

Let us choose affine coordinates (x,y) on $\mathbb{P}^1 \times \mathbb{P}^1 - A$: then g belongs to the group generated by the involution g_3 such that $g_3(x,y) = (y,x)$, and by the two involutions g_1, g_2 such that $g_1(x,y) = (1-x,y)$, $g_2(x,y) = (x,1-y)$.

Let $r = g_3g_1$: then r has period 4; if we set $s = g_3$, then $s^2 = 1$, $r^4 = 1$, $sr^3 = rs = g_2$, $r^2 = g_1g_2$, and our group is the dihedral group D_4.
We can thus reformulate our discussion with the following

Theorem 1.19. $M_{4,1}$ is the quotient of $\mathbb{P}^{10}$ by a suitable action of the dihedral group D_4.

For the geometrical construction underlying theorem C, consider again a non-hyperelliptic curve $C \subset \mathbb{P}^1 \times \mathbb{P}^1 = Q$. In this picture we have in Q a family of lines of the form $\mathbb{P}^1 \times \{a\}$, $a \in \mathbb{P}^1$, and another of the form $\{b\} \times \mathbb{P}^1$, $b \in \mathbb{P}^1$, which we visualize as being orthogonal to the first.

Definition 1.20. A rectangle R in Q is the union of four distinct lines in Q, of the form $R = (\mathbb{P}^1 \times \{b\}) \cup (\mathbb{P}^1 \times \{b'\}) \cup (\{a\} \times \mathbb{P}^1) \cup (\{a'\} \times \mathbb{P}^1)$. Its vertices are the four points (a,b), (a,b'), (a',b), $(a'b')$ and if they all belong to C we shall say that R is inscribed into C.

Theorem 1.21. A general curve C of genus 4 admits 6 inscribed rectangles (lying in the unique quadric Q containing the canonical image of C).

Proof. Consider C^4 and the four projections $f_i : C^4 \to C$ $(i=1,\ldots 4)$ on the four factors of the product. Let moreover $p, p' : Q \to \mathbb{P}^1$ be the two natural projections: they define two divisors of degree 3 on C, which we denote, respectively, by D and D'.

Let D_i be the divisor on C^4 such that $D_i = f_i^*(D)$ (resp. $D_i' = f_i^*(D')$); let moreover $\Delta_{ij} \subset C^4$ be $\{(y_1,y_2,y_3,y_4)\ \ y_i = y_j\}$

Consider in C^4 the subvariety $W = \{(y_1,y_2,y_3,y_4) \mid p(y_1) = p(y_2),$ $p(y_3) = p(y_4),\ p'(y_1) = p'(y_4),\ p'(y_2) = p'(y_3)\}$.

Given an inscribed rectangle R and a vertex x of R one determines a unique point $y = (y_1,y_2,y_3,y_4)$ in W with $y_1 = x$, and such that $y \in W - \bigcup_{i \leq j} \Delta_{ij}$.

Conversely, if $y \in W - \Delta_{12} - \Delta_{34} - \Delta_{14} - \Delta_{23}$, then also $y_1 \neq y_3$ since otherwise $p'(y_1) = p'(y_3) = p'(y_2)$, and, since $p(y_1) = p(y_2)$, one would have $y_1 = y_2$; analogously one has $y_2 \neq y_4$.
Therefore the points of $W - \Delta_{12} - \Delta_{34} - \Delta_{14} - \Delta_{23}$ are in a bijection with the pairs (R,x) where R is a rectangle inscribed into C, x is a vertex of R.
Now the above mentioned set is the complete intersection of four divisors.
In fact, consider in C^2 the divisor $B = \{(y_1,y_2) \mid p(y_1) = p(y_2)\}$. $B = \Delta + \Gamma$, where Δ is the diagonal of $C \times C$, and Γ is smooth away from Δ since p is a covering of degree equal to three.
B is the pull back of the diagonal in $\mathbb{P}^1 \times \mathbb{P}^1$ under the morphism $p \times p : C^2 \to (\mathbb{P}^1)^2$, therefore its class as a divisor on C^2, using our previous notations ($f_i : C^2 \to C$, $i=1,2$, being the two projections), is just $D_1 + D_2$. Since C has genus four $\Delta^2 = -6$, moreover $B \cdot \Delta = 6$, so that $\Gamma \cdot \Delta = 12$.

Consider the monodromy of $p: C \to \mathbb{P}^1$: if C is general, then p has only ordinary ramification, i.e.

a) Γ and Δ intersect transversally

b) the monodromy of p is generated by transpositions

Since C is connected b) implies that the monodromy is the full symmetric group, hence, in general, Γ is smooth, irreducible, transversal to Δ in the points corresponding to the ramification points of p.

Considering the projection p' instead of p, we define analogously $\Gamma' \subset C^2$.

Let $\Gamma_{ij} = (f_i \times f_j)^*(\Gamma)$, $\Gamma'_{hk} = (f_h \times f_k)^*(\Gamma')$.

Then we claim that $W - \Delta_{12} - \Delta_{34} - \Delta_{14} - \Delta_{23} = \Gamma_{12} \cap \Gamma_{34} \cap \Gamma'_{14} \cap \Gamma'_{23}$, and the intersection is transversal, for C general.

In fact, if $(y_1, y_2) \in \Delta_{12} \cap \Gamma_{12}$, then $y_1 = y_2$ and y_1 is a ramification point of p: since $y = (y_1, \ldots y_4) \in W$ it follows that $y_3 = y_4$, hence y_3 is a second ramification point of p, and $p'(y_1) = p'(y_3)$.

It is easy to see that curves C of type $(3,3)$ in Q such that the above situation can hold form a proper subvariety in the linear system $|\mathcal{O}_Q(3,3)|$

To show that $\Gamma_{12} \cap \Gamma_{34} \cap \Gamma'_{14} \cap \Gamma'_{23}$ gives a transversal intersection for C general, we consider the variety $\Lambda \subset |\mathcal{O}_Q(3,3)| \times Q^4$ defined by

$$\Lambda = \{(C, y_1, y_2, y_3, y_4) \mid y_i \in C,\ i=1,\ldots 4,\ p(y_1) = p(y_2),\ p(y_3) = p(y_4),$$
$$p'(y_2) = p'(y_3),\ p'(y_1) = p'(y_4)\}$$

Λ is of dimension 15 and smooth at the general point, hence our assertion is proven if the projection of Λ on $|\mathcal{O}_Q(3,3)|$ is surjective: but if this were not the case, for C general, $\Gamma_{12} \cap \Gamma_{34} \cap \Gamma'_{14} \cap \Gamma'_{23}$ would be empty.

Finally we compute: $\Gamma_{12} \cdot \Gamma_{34} \cdot \Gamma'_{14} \cdot \Gamma'_{23} = (D_1 + D_2 - \Delta_{12}) \cdot (D_3 + D_4 - \Delta_{34}) \cdot (D'_1 + D'_4 - \Delta_{14}) \cdot$ $\cdot (D'_2 + D'_3 - \Delta_{23}) = 2 \cdot 3^4 - 2 \cdot 3^3 \cdot 4 + 2 \cdot 3^2 \cdot 6 - 2 \cdot 3 \cdot 4 - 6 = 3^3 \cdot 2 - 30 = 24$; in fact $\Delta_{12} \cdot \Delta_{34} \cdot \Delta_{14} \cdot \Delta_{23}$ equals the self-intersection of Δ in $C \times C$.

Q.E.D.

Theorem C is now a straightforward consequence of theorem 1.21.

Namely, consider in $Q = \mathbb{P}^1 \times \mathbb{P}^1$ the following set of six points:

$M' = \{(\infty,\infty), (0,0), (0,\infty), (\infty,0), (1,\infty), (\infty,1)\}$.

Let $|G'|$ be the linear system $|\mathcal{I}_{M'}(3,3)|$: we can choose affine coordinates (x,y) on $\mathbb{P}^1 \times \mathbb{P}^1 - \{\infty\} \times \mathbb{P}^1 - \mathbb{P}^1 \times \{\infty\}$. Then $|G'|$ is the projective space associated with the vector space U spanned by the monomials

$$\begin{cases} x,\ x^2,\ x^2 y,\ x^3 y(1-y) \\ y,\ y^2,\ y^2 x,\ y^3 x(1-x) \\ xy,\ x^2 y^2 \end{cases}$$

These monomials are permuted by the action on $|G'|$ induced by the automorphism $s:Q \to Q$ such that $s(x,y) = (y,x)$. It is then obvious that $|G'|_{/s} = \mathbb{P}(U)/_s$ is birational to $\mathbb{P}^9$. We conclude then this paragraph with

<u>Theorem C</u>. M_4 has covering of degree 24 by a rational variety. More precisely, the rational map of $|G'| \to M_4$ is a covering of degree 48 which factors through the action of s on $|G'|$, and a general point of $|G'|_{/s}$ corresponds to the datum of a triple (C,R,p) where C is a curve of genus 4, R is a rectangle inscribed into C, p is a vertex of R.

<u>Proof</u>. To a curve $C \in |G'|$ we associate the triple $(C, (\mathbb{P}^1 \times \{0,\infty\} \cup (\{0,\infty\} \times \mathbb{P}^1), (\infty,\infty))$.

Assume now that C, C' give isomorphic triples: then there exists $g \in \mathrm{Aut}(Q)$ such that $g(\infty,\infty) = (\infty,\infty)$, $g(C) = C'$, and for the rectangle $R = (\mathbb{P}^1 \times \{0,\infty\}) \cup (\{0,\infty\} \times \mathbb{P}^1)$ one has $g(R) = R$. In particular, $g(M') = M'$, so that necessarily g is either the identity or the involution s. The fact the degree of the rational map of $|G'|$ onto M_4 is 48 follows immediately from theorem 1.21.

Q.E.D.

§2. RATIONALITY OF THE INVARIANT SUBFIELDS.

Before turning to prove the rationality of R_4, we first state a more general auxiliary result.

Let V_n be the standard permutation representation of the symmetric group $\underline{S}_n$, V_n^m the direct sum of m copies of V_n. Then the field of rational functions on V_n^m, $k(V_n^m)$, can be written as $k(x_{11},\dots x_{1n}, x_{21},\dots x_{2n},\dots,x_{m1},\dots x_{mn})$ and a permutation τ acts on x_{ij} by sending it to $x_{i\tau(j)}$. Consider the following invariant rational functions, where σ_i denotes the i-th elementary symmetric function, and a variable with a cap has to be omitted:

$$(2.1)\quad \begin{cases} \sigma_i' = \sigma_i(x_{11},\dots x_{1n}) & i = 1,\dots n \\ \sigma_1^{(h)} = \sum_{j=1}^{n} x_{hj}x_{1j} & h = 2,\dots m \\ \sigma_i^{(h)} = \sum_{j=1}^{n} x_{hj}\,\sigma_{i-1}(x_{11},\dots,\hat{x}_{1j},\dots x_{1n}) & h = 2,\dots m \\ & i = 2,\ \dots m. \end{cases}$$

<u>Lemma 2.2</u>. The invariant subfield $k(V_n^m)^{\underline{S}_n}$ is a rational field: more precisely the nm functions given by 2.1 form a basis of the purely transcendental extension over k.

<u>Proof</u>. $\sigma', \sigma^{(2)}, \dots, \sigma^{(m)}$ determine a morphism $\psi: V_n^m \to (\mathbb{A}^n)^m$ and to prove that ψ induces a birational map of $V_n^m/\underline{S}_n$ onto the affine space $(\mathbb{A}^n)^m$ it is enough to prove that on a Zariski open set of V_n^m $\psi(x) = \psi(y)$ if and only if there exists $\tau \in \underline{S}_n$ such that $\tau(x) = y$.
The "if" part being obvious, let's assume that $\psi(x) = \psi(y)$: then, in particular, $\sigma'(x) = \sigma'(y)$.

By virtue of the fundamental theorem on symmetric functions, we can assume, acting on y by a suitable $\tau \in \underline{S}_n$, that $x_{1j} = y_{1j}$ for $j = 1,\dots n$.

Let us set for convenience $z_j = x_{1j}$ $(j=1,\dots,n)$. Then the variables x_{hj}, y_{hj} $(h=2,\dots m,\ j=1,\dots n)$ are solutions, by 2.1, of the same system of $n(m-1)$ linear non-homogeneous equations, hence they are equal if the determinant of the system is non-zero.

The system being given by the matrix

$$(2.3)_n\quad \begin{pmatrix} z_1, & \dots, z_n \\ \sigma_1(z_2,\dots z_n) & \dots, \sigma_1(z_1,\dots,z_{n-1}) \\ \vdots & \vdots \\ \sigma_{n-1}(z_2,\dots z_n) & \dots, \sigma_{n-1}(z_1,\dots,z_{n-1}) \end{pmatrix}$$

it suffices to verify that the determinant of the matrix $(2.3)_n$ is not identically zero. We prove this by induction on n, since for $n = 2$ we get $\det\begin{pmatrix} z_1 & z_2 \\ z_2 & z_1 \end{pmatrix} = z_1^2 - z_2^2$. For bigger n, the determinant of $(2.3)_n$ modulo z_n, is given, up to sign, by the product of $z_1 \dots z_{n-1} = \sigma_{n-1}(z_1, \dots z_{n-1})$ times the determinant of $(2.3)_{n-1}$.

Q.E.D.

Theorem A. R_4 is a rational variety.

Proof. In view of theorem 1.18 we have to show the rationality of $\mathbb{P}(\mathrm{Sym}^2(V_4))/\underline{S}_4$. We use here the fact that $\underline{S}_4$ has a normal subgroup $G \cong (\mathbb{Z}/2)^2$ given by the double cycles in $\underline{S}_4$; the quotient $\underline{S}_4/G$ is isomorphic to $\underline{S}_3$ and in this way any representation of $\underline{S}_3$ induces canonically a representation of $\underline{S}_4$ that we shall denote by the same symbol.

Since the action of $\underline{S}_4$ on $\mathrm{Sym}^2(V_4)$ is linear, it is clearly sufficient to prove the rationality of the quotient $\mathrm{Sym}^2(V_4)/\underline{S}_4$.

We subdivide the proof in four steps, noticing that we have the following chain of inclusions

$$(2.4) \qquad k(\mathrm{Sym}^2(V_4)) \supset k(\mathrm{Sym}^2(V_4))^G \supset k(\mathrm{Sym}^2(V_4))^{\underline{S}_4} = \left(k(\mathrm{Sym}^2(V_4))^G\right)^{\underline{S}_3}.$$

Let W_4 be the irreducible $\underline{S}_4$-submodule of V_4 generated by $x_1 - x_2$, $x_2 - x_3$, $x_3 - x_4$: $V_4 = \mathbb{1} \oplus W_4$, $\mathbb{1}$ being the trivial one dimensional representation spanned by $\sigma_1(x_1, \dots x_4)$.

Step I. $\mathrm{Sym}^2(V_4) \cong \mathbb{1} \oplus W_4^2 \oplus V_3$

Proof. $\mathrm{Sym}^2(V_4) \cong V_4' \oplus V_3' \oplus W_4'$ where V_4' is spanned by $x_1^2, x_2^2, x_3^2, x_4^2$, V_3' is spanned by $y_1 = x_1x_2 + x_3x_4$, $y_2 = x_1x_3 + x_2x_4$, $y_3 = x_1x_4 + x_2x_3$, W_4' is spanned by $w_1 = x_1x_2 - x_3x_4$, $w_2 = x_1x_3 - x_2x_4$, $w_3 = x_1x_4 - x_2x_3$.

V_4' is clearly isomorphic to V_4; also, since G acts trivially on V_3', V_3' is induced by a representation of $\underline{S}_3$.

V_3' has as basis three vectors corresponding to the three non-trivial double cycles of $\underline{S}_4$, and the action of $\underline{S}_4$ on the basis is given by conjugation in $\underline{S}_4$ (G acts trivially being abelian).

Observing that the transposition (1,4) permutes y_1 with y_2 and leaves y_3 fixed, (1,2) leaves y_1 fixed and permutes y_2 with y_3, we conclude that V_3' is isomorphic to V_3.

On W'_4 we have the following actions:

$$(12)(34) \text{ acts by } \begin{cases} w_1 \mapsto w_1 \\ w_2 \mapsto -w_2 \\ w_3 \mapsto -w_3 \end{cases}, \quad (12) \text{ by } \begin{cases} w_1 \mapsto w_1 \\ w_2 \mapsto -w_3 \\ w_3 \mapsto -w_2 \end{cases}$$

$$(123) \text{ by } \begin{cases} w_1 \mapsto w_3 \\ w_2 \mapsto w_1 \\ w_3 \mapsto -w_2 \end{cases}, \quad (1234) \text{ by } \begin{cases} w_1 \mapsto -w_3 \\ w_2 \mapsto -w_2 \\ w_3 \mapsto w_1 \end{cases}$$

Let χ' be the character of W'_4: the character of W_4 equals the character χ of V_4 minus 1, hence we conclude that $\chi - 1 = \chi'$ by computing explicitly the table of characters

Conjugacy classes	id	(12)(34)	(12)	(123)	(1234)
χ'	3	-1	1	0	-1
χ	4	0	2	1	0

If the characteristic of k is different from 2,3, this implies that $W'_4 \cong W_4$; in characteristic 3, this is also true, because both representations are irreducible: in fact (cf. [15], pag. 155) their modular characters are indecomposable. □

To compute $k(\mathrm{Sym}^2(V_4))^G$, in view of step I, suffices

<u>Step II</u>. $k(W_4^2)^G = L$, if $w_1, w_2, w_3, w'_1, w'_2, w'_3$ are coordinates on $W_4 \oplus W_4$, is generated by

(2.5) $\quad w_1^2, w_2^2, w_1w_2w_3, w_iw'_i \quad (i=1,2,3)$

<u>Proof</u>. The six given functions are G-invariant, and $k(W_4^2) = L(w_1,w_2)$, so we have an extension of degree 4 and L is the whole subfield of G-invariants. □

<u>Step III</u>. Let F be the subfield of L generated by $w_1^2, w_2^2, w_3^2, w_iw'_i$. Then L

is a quadratic extension of F given by $F(t)$, where $t = w_1w_2w_3$. $F = k(V_3^2)$ as a representation of $\underline{S}_3 = \underline{S}_4/G$.

Proof. Clearly $t \notin F$, $t^2 = (w_1^2)(w_2^2)(w_3^2)$. Also the action of $\underline{S}_4$ on V_3 differs from the one of $\underline{S}_4$ on W_3 only up to sign, i.e., as it is easy to verify, $\underline{S}_4$ acts by permuting the basis given by y_1, y_2, y_3, and if $\tau(y_i) = y_j$, then $\tau(w_i) = \pm w_j$, hence $\tau(w_i^2) = w_j^2$, $\tau(w_iw_i') = w_jw_j'$.

□

Step IV. Let M be the field generated by w_i^2, $w_i'w_i$, y_i $(i=1,2,3)$. M is a purely transcendental extension of k, $M \cong k(V_3^3)$, $k(\mathrm{Sym}^2V_4)^{\underline{S}_4} = M^{\underline{S}_3}(t,\sigma)$, where $\sigma = \sigma_1(x_1,\dots x_4)$, $t = w_1w_2w_3$.

Proof. $k(\mathrm{Sym}^2V_4)^{\underline{S}_4} = (M(t,\sigma))^{\underline{S}_3}$, but σ is an invariant for $\underline{S}_4$ from the very beginning, while t is an $\underline{S}_3$ invariant by the formulas written in step I. That M is isomorphic to $k(V_3^3)$ follows by step III.

□

End of the proof. $k(\mathbb{P}(\mathrm{Sym}^2V_4))^{\underline{S}_4} = M^{\underline{S}_3}(t)$. But, by lemma 2.1, $M^{\underline{S}_3}$ is a rational field with basis of transcendency $\sigma_1', \sigma_2', \sigma_3', \sigma_1^{(2)}, \sigma_2^{(2)}, \sigma_3^{(2)}, \sigma_1^{(3)}, \sigma_2^{(3)}, \sigma_3^{(3)}$. We conclude observing that $\sigma_i' = \sigma_i(w_1^2,w_2^2,w_3^2)$, hence $t^2 = \sigma_3'$.

Q.E.D.

Theorem B. $M_{4,1}$ is a rational variety.

Proof. By theorem 1.19 and the arguments preceding it we have a linear representation $\rho:D_4 \to \mathrm{Aut}(U)$ where U is 11-dimensional, and we know that $M_{4,1}$ is birational to $\mathbb{P}(U)/D_4$.
U has a basis given by the polynomials $1, x, x^2, y, y^2, xy, x^2y, xy^2, x^2y^2, xy^3(1-x), x^3y(1-y)$, and, if r, s are the generators of D_4, such that $s^2 = r^4 = 1$, $sr^3 = rs$,

(2.6) $\quad s(x,y) = (y,x), \quad r(x,y) = (y,1-x).$

To decompose U as a direct sum of irreducibles, since D_4 has order 8 and we assume $\mathrm{char}(k) \neq 2$, we compute the character χ of ρ.
For s we observe that $\rho(s)$ permutes the elements of the basis, leaving $1, xy, x^2y^2$ fixed: hence $\chi(s) = 3$.
For sr, $sr(x,y) = (1-x,y)$ and choosing for U the new basis $x, (1-x), x(1-x), yx, y(1-x), yx(1-x), y^2x, y^2(1-x), y^2x(1-x), y^3x(1-x), x^3y(1-y)$, we see that the

trace of $\rho(sr)$ is 3 since sr permutes the first 10 elements of the basis, leaving 4 of them fixed, while $sr(x^3y(1-y)) = (1-x)^3(1-y)y = -x^3y(1-y)$ + terms of lower degree in x.
For $\chi(r)$, $x, x^2, y, y^2, x^2y, xy^2, xy^3(1-x), x^3y(1-y)$ are easily seen to give a zero contribution (by degree considerations), while 1 is invariant, $xy \mapsto y - xy$, $x^2y^2 \mapsto y^2 - 2xy^2 + x^2y^2$, therefore $\chi(r) = 1$.
Since $r^2(x,y) = (1-x, 1-y)$, the trace of $\rho(r^2)$ is easily seen (in the first given basis) to be equal to $1 - 1 + 1 - 1 + 1 + 1 - 1 - 1 + 1 - 1 - 1 = -1$.
We now put together the character χ of ρ and the characters of the irreducible representations of D_4.

(2.7)

Conjugacy classes		1	$\{r, r^3\}$	$\{s, sr^2\}$	sr, sr^3	r^2
characters	ψ_1	1	1	1	1	1
	ψ_2	1	1	-1	-1	1
	ψ_3	1	-1	1	-1	1
	ψ_4	1	-1	-1	1	1
	χ'	2	0	0	0	-2
	χ	11	1	3	3	-1

Since χ', and the ψ_i's are an orthogonal basis for the space of class functions, by computing scalar products we obtain that $\chi = 3\psi_1 + \psi_3 + \psi_4 + 3\chi'$. Now ψ_1 is the trivial representation, hence we conclude:

<u>Step I.</u> $k(\mathbb{P}(U)/D_4)$ is a purely transcendental extension of degree 2 of the invariant subfield $k(V)^{D_4}$, where V is the representation with character $3\chi' + \psi_3 + \psi_4$.

<u>Step II.</u> The cyclic subgroup generated by r is normal, hence $k(V)^{D_4} = (k(V)^r)^{\mathbb{Z}/2}$.

Now, if i is a square root of -1, then the representation λ corresponding to χ' is given by

$$\lambda(r) = \begin{pmatrix} i & 0 \\ 0 & -i \end{pmatrix} \quad , \quad \lambda(s) = \begin{pmatrix} 0 & 1 \\ 1 & 0 \end{pmatrix} . \tag{2.8}$$

Therefore we can choose coordinates x_j, y_j $(j=1,2,3)$, z_1, z_2 on V such that r acts by $x_j \mapsto ix_j$, $y_j \mapsto iy_j$, $z_h \mapsto -z_h$ while s acts by permuting x_j with y_j, while $s(z_1) = z_1$, $s(z_2) = -z_2$.

Step III. $k(V)^r$ is a purely transcendental extension of k, K, generated by x_1^4, x_1/x_2, x_1/x_3, x_1y_1, y_1/y_2, y_1/y_3, $z_1(x_1^2+y_1^2)$, $z_2(x_1^2+y_1^2)$.

Proof. $K \subset k(V)^r$, and clearly $k(V) = K(x_1)$, but $x_1^4 \in K$, so we have equality. Unfortunately in this way the action of s is not linear any more: to avoid this we replace first in the basis x_1^4 by $u = x_1^2/y_1^2 = x_1^4/x_1^2y_1^2$.

Then $s(u) = 1/u$: finally we replace u by $(u-1)/(u+1) = w$ so that $s(w) = -w$.

End of the proof. In this way we have a linear action of $\mathbb{Z}/2$ on an 8-dimensional vector space, and with 4 eigenvalues equal to $(+1)$, 4 equal to (-1). The quotient is obviously rational.

Q.E.D.

REFERENCES

[1] Arbarello, E. - Sernesi, E.: The equation of a plane curve, Duke Math.J. 46, 2(1979), 469-485.

[2] Catanese, F.: Babbage's conjecture, contact of surfaces, symmetric determinantal varieties and applications, Inv.Math. 63(1981), 433-465.

[3] Catanese, F. - Ceresa, G.: Constructing sextic surfaces with a given number d of nodes, J.Pure & Appl.Alg. 23(1982), 1-12.

[4] Clemens, H.: Double solids, Advances in Mathematics, to appear.

[5] Coble, A.B.: Algebraic geometry and theta functions, Coll.Publ. of the A.M.S. vol.10 (1929) (reprint 1961).

[6] Donagi, R.: The unirationality of A_5, to appear.

[7] Harris, J. - Mumford, D.: On the Kodaira dimension of the moduli space of curves, Inv.Math.67, 1(1982) 23-86.

[8] Igusa, J.I.: Arithmetic variety of moduli for genus two, Annals of Math. 72(1960), 612-649.

[9] Mori, S.: The uniruledness of M_{11}, to appear.

[10] Mumford, D.: Geometric invariant theory, Springer Verlag (1965).

[11] Mumford, D.: Varieties defined by quadratic equations, in "Questions on algebraic varieties", C.I.M.E. Varenna, 1969, Cremonese.

[12] Mumford, D.: Prym varieties I, in "Contributions to Analysis", Academic Press (1974), 325-350.

[13] Recillas, S.: La variedad de los modulos de curvas de genero 4 es unirracional, Ann.Soc.Mat. Mexicana (1971).

[14] Sernesi, E.: L'unirazionalitá della varietá dei moduli delle curve di genere dodici, Ann. Scuola Norm.Sup. Pisa, 8(1981), 405-439.

[15] Serre, J.P.: Linear representations of finite groups, G.T.M. 42, Springer Verlag (1977).

[16] Severi, F.: Vorlesungen über Algebraïschen Geometrie, Teubner, Leipzig (1921).

[17] Van der Geer, G.: On the geometry of a Siegel modular threefold, Math.Ann. 260, (1982) 317-350.

[18] Wirtinger, W.: Untersuchungen über Thetafunktionen, Teubner, Leipzig (1895)

A CONSTRUCTION OF SPECIAL SPACE CURVES

D. Gieseker*
University of California
Los Angeles, California 90024

§1. We will work over a fixed algebraically closed field of characteristic zero. A curve $C \subseteq \mathbb{P}^r$ is said to be non-degenerate if it does not lie in a hyperplane. Let d' and g' be positive integers, and let $r \geq 3$ be an integer.

PROPOSITION 1.1. *If* $d' \geq (r+3) + \left(\frac{r-1}{r+1}\right)g'$, *then there are non-degenerate smooth curves of genus* g' *and degree* d' *in* $\mathbb{P}^r$.

For $r = 3$, we will establish a somewhat stronger result:

PROPOSITION 1.2. *If* $d' \geq \frac{1}{3}g' + 10$, *then there are non-degenerate smooth curves of genus* g' *and degree* d' *in* $\mathbb{P}^3$.

After completing this paper, I learned that Gruson and Peskine have constructed smooth curves in $\mathbb{P}^3$ of degree d' and genus g' if $g' \leq \frac{1}{6}d(d-3)+1$. So Proposition 1.2 is a very special case of their result. However, their construction is quite different from the one given here.

To establish Propositions 1.1 and 1.2, we produce certain stable curves of genus g' and degree d' in $\mathbb{P}^r$. Then by an argument involving counting constants, we show that these curves deform to smooth curves.

We will illustrate the proof of Proposition 1.1 by considering the case $r = 3$ and constructing certain space curves. We begin with a smooth nondegenerate curve $C \subseteq \mathbb{P}^3$ of genus g and degree d. We suppose that C has no automorphisms, that $H^1(C, \mathcal{O}_C(1)) = 0$, and that for a generic hyperplane $H \subseteq \mathbb{P}^3$, that no conic in H contains $H \cap C$. Now choose hyperplanes $H_1, \ldots, H_k$ in $\mathbb{P}^3$ generically and choose conics R_i in H_i so that R_i contains five points of $H_i \cap C$. Let $D = C \cup R_1 \cup \cdots \cup R_k$. We will show that D is a stable curve of genus $g' = g + 4k$ and degree $d' = d + 2k$. Let $L = \mathcal{O}_D(1)$.

Now D deforms in $\mathbb{P}^3$ to a smooth curve if $2k \geq d - g + 3$. This can be seen intuitively from the following non-rigorous dimension count. A stable curve X will be called good if it has $(k+1)$ smooth components $C_0, C_1, \ldots, C_k$, where C_0 is C and $C_1, \ldots, C_k$ are non-singular rational, and C_i meets C in five nodes, but $C_i \cap C_j = \emptyset$ if $i \neq j$. Let

$$T = \left\{ (X,M) \;\middle|\; \begin{array}{l} X \text{ is a good stable curve,} \\ M \text{ is a line bundle on } X, \\ M_{C_0} \cong L \text{ and } M \text{ has degree } 2 \\ \qquad \text{on each } C_i \text{ for } i \geq 1. \end{array} \right\}$$

*Partially supported by NSF Grant MCS 79-3171

The pair (D,L) constructed above is good. How many moduli does T have? A good stable curve is determined by $5k$ points in C and k sets of five points in $\mathbb{P}^1$ determined up to projective equivalence in $\mathbb{P}^1$. Thus there are $7k$ moduli for the good stable curves. Now if X is fixed, then the set of $(X,M) \in T$ has $4k$ moduli. Thus T has $11k$ moduli.

Now let

$$T_3 = \{(X,M) \in T \mid h^0(X,M) = 4\} .$$

We will show $(D,L) \in T_3$. How many moduli does T_3 have? Since L is very ample, the generic element of T_3 is very ample. Let G be the Grassmannian of four dimensional subspaces of $H^0(C,L)$. Now each (X,M) in T_3 determines a four dimensional subspace $H^0(X,M)$ in $H^0(C,L)$. Let $\varphi(X,M) \in G$ denote this subspace. Now G has dimension $4(d - g - 3)$. Now suppose $\varphi(X,M) = \varphi(D,L)$. Then we may assume that $X \subseteq \mathbb{P}^r$ and that C is a component of X. Further, $M \cong \mathcal{O}_X(1)$. If the C_i are the rational components of X, then C_i lies in a hyperplane, since C_i has degree two. Now suppose that $C \cap R_i$ and $C \cap C_i$ have three points in common for each i. Then the hyperplane containing R_i and C_i must be the same. Hence the points of $C \cap C_i$ are determined up to finitely many choices if three points of $C \cap C_i$ coincide with three points of $C \cap R_i$. Further, if $C \cap C_i = C \cap R_i$, then $C_i = R_i$, since the conic through five points in a general position is unique. Also in this case L and M are isomorphic, since they are both $\mathcal{O}_D(1)$. Thus T_3 depends on at most $4(d - g - 3) + 3k$ parameters. Let $\rho = 4(3 + g' - d')$. A short computation shows that $\operatorname{codim}_T T_3 \geq \rho$.

Now let S be the set of all (X,L), where X is a stable curve of genus g' and L is a line bundle of degree d' on X. Let S_s be the set of all $(X,M) \in S$ with X singular, and let S_3 be the set of all (X,M) with $h^0(X,M) = 4$. Then $S_3 \cap T = T_3$. Further, it is easy to see that $\operatorname{codim}_S S_3 \leq \rho$. Using the theory of the moduli of stable curves, one sees T is contained in every component of S_s which meets T. But if $S_3 \subseteq S_s$, then $\operatorname{codim}_{S_s} S_3 \leq \rho - 1$. But we have $\operatorname{codim}_T T_3 \geq \rho$. So $S_3 \not\subseteq S_s$. Now since L is very ample on D, there are curves of degree $d + 2k$ and genus $g + 4k$ in $\mathbb{P}^3$.

The idea behind the proof of Proposition 1.2 is similar to the above, except that we draw cubics through nine points of $H_i \cap C$.

I wish to thank Joe Harris for suggesting the monodromy arguments used in this paper.

§2. Let $\pi : X \to S$ be a flat family of stable curves of genus g parametrized by some variety S. If $s \in S$, we let $X_s = \pi^{-1}(s)$. We say $\pi : X \to S$ is equinodal if all the X_s have the same number of nodes.

Let $\mathcal{L}$ be a line bundle on X and let $\mathcal{L}_s$ be the induced line bundle on X_s, and suppose $\mathcal{L}_s$ has degree d. Let S_r be the locally closed subvariety of S:

$$S_r = \{s \in S \mid h^0(X_s,L_s) = r+1\} .$$

Let $\rho = \rho(g,d,r) = (r+1)(g-d+r)$.

DEFINITION 2.1. Let $s \in S_r$. We say $\mathcal{L}$ satisfies the Brill-Noether condition at s if either $h^1(X_s,L_s) = 0$ or the codimension of every component of S_r containing s is ρ.

PROPOSITION 2.2. _Suppose that_ $\pi : X \to S$ _is equinodal, that_ $\mathcal{L}$ _satisfies the Brill-Noether condition at_ s _and that_ $\mathcal{L}_s$ _is very ample. Then there are smooth non-degenerate curves of degree_ d _and genus_ g _in_ $\mathbb{P}^r$, _where_ $r+1 = h^0(\mathcal{L}_s)$.

Proof. To establish the proposition, we may replace S by any neighborhood of s. Let ω be the relative dualizing sheaf for $\pi : X \to S$. By shrinking S, we may assume $\pi_*(\omega^{\otimes 3} \otimes \mathcal{L})$ is free on S. Let $\mathcal{H}$ be the Hilbert scheme of all curves C of degree $d + 3(2g-2)$ and genus g in $\mathbb{P}^n$, with $h^1(C,\mathcal{O}(1)) = 0$, where $n = d + 3(2g-2) + 1 - g$. Let $p : \mathcal{C} \to \mathcal{H}$ be the tautological curve over $\mathcal{H}$, and let $\mathfrak{m}$ be the tautological line bundle on $\mathcal{C}$. Choosing a basis of $\pi_*(\omega^{\otimes 3} \otimes \mathcal{L})$, we obtain a morphism $\Phi : S \to \mathcal{H}$ so that $X \cong \mathcal{C} \times_{\mathcal{H}} S$ and $L \otimes \omega^{\otimes 3}$ is isomorphic to $\Phi^*(\mathfrak{m})$. We also denote the dualizing sheaf of $\mathcal{C} \to \mathcal{H}$ by ω. Let $\mathfrak{n}$ be $\mathfrak{m} \otimes \omega^{\otimes -3}$. Then $\Phi^*(\mathfrak{n})$ is just $\mathcal{L}$. Further, if $\mathcal{H}_r = \{h \in \mathcal{H} \mid h^0(C_h,\mathfrak{n}_h) = r+1\}$, then $\Phi^{-1}(\mathcal{H}_r) = S_r$. Now let

$$T = \{h \in \mathcal{H}_r \mid C_h \text{ is singular}\} .$$

Suppose C_h is stable. We recall the local structure of T at h (cf. [1]). Namely, $\mathcal{H}$ is smooth, and T has normal crossing singularities at h, and locally, the points of T with multiplicity k correspond to stable curves with k nodes. In particular since $\pi : X \to S$ is equinodal, we see $\Phi(S)$ is contained in any component T_i of T which intersects $\Phi(S)$.

Now suppose $\mathcal{H}_r$ were contained in T. Let W be a component of $\mathcal{H}_r$ which contains $\Phi(S)$. We claim that $\operatorname{codim}_{\mathcal{H}} W \le \rho$. First let us define a functor F from the category of $\mathcal{H}$ schemes of finite type to the category of sets. For each $\mathcal{H}$ scheme $T' \to \mathcal{H}$, let

$$F(T') = \left\{ D \;\middle|\; \begin{array}{l} D \text{ an effective relative Cartier divisor on} \\ S \times_{\mathcal{H}} \mathcal{C} \text{ so that } \mathcal{O}(D) \text{ is locally isomorphic} \\ \text{over } T' \text{ to the pullback of } \mathfrak{n} \end{array} \right\} .$$

Using Grothendieck's Quot scheme, there is an $\mathcal{H}$ scheme $\mathfrak{W}$ and a relative Cartier divisor $\mathfrak{D}$ on $\mathfrak{W} \times_{\mathcal{H}} \mathcal{C} = \mathcal{C}'$ which represents F. Thus a closed point w_0 of $\mathfrak{W}$ consists of a closed point h_0 of $\mathcal{H}$ and a Cartier divisor D_0 on C_{h_0} so that $\mathcal{O}(D_0)$ is isomorphic to $\mathfrak{n}_{h_0}$.

Let $\mathcal{H}_{n.s.}$ denote the open set of $\mathcal{H}$ over which $\pi : \mathcal{C} \to \mathcal{H}$ is smooth and let $\mathfrak{W}_{n.s}$ be the inverse image of $\mathcal{H}_{n.s}$ in $\mathfrak{W}$. Note first that $\mathfrak{W}_{n.s.}$ is dense in $\mathfrak{W}$.

Indeed, let $w_0 \in \mathfrak{w}$ and let C_0 and D_0 be the corresponding curve and divisor. Assume first that the support of D_0 is disjoint from the singular locus of C_0. We can find flat family of curves $\pi : C \to S'$ parametrized by a smooth curve S' and a relative Cartier divisor D on C so that for some closed point $s_0 \in S_0$, $C_{s_0} = C_0$ and $D_{s_0} = D$ and so that the generic fiber of π is smooth. Passing to a neighborhood of s_0, we may assume that $h^1(C_s, \mathcal{O}(D) \otimes \omega^{\otimes 3}) = 0$, since an easy duality argument shows that $h^1(\mathcal{O}(D_0) \otimes \omega^{\otimes 3}) = 0$. Choose a basis of $\pi_*(\mathcal{O}(D) \otimes \omega^{\otimes 3})$ so that the induced map $\psi : S' \to \mathfrak{H}$ passes through the image of w_0 in $\mathfrak{H}$. Then D determines a map $\psi' : S \to \mathfrak{w}$ lifting ψ. Thus $\psi'(s_0) = w_0$. Thus $w_0 \in \overline{\mathfrak{w}_{n.s.}}$.

Suppose D_0 contains singular points of C_0. Deforming D_0 on C_0 is a local question, so we can deform D_0 to a divisor whose support misses the singular points of C_0. The previous argument shows $w_0 \in \overline{\mathfrak{w}_{n.s.}}$.

We write $\dim X \geq r$ to mean any component has dimension $\geq r$. We next claim that

(2.2.1) $$\dim \mathfrak{w} \geq \dim \mathfrak{H} + d - g .$$

Indeed, it suffices to establish that $\dim \mathfrak{w}_{n.s.} \geq \dim \mathfrak{H}_{n.s.} + d - g$. First note that the fiber of $\mathfrak{w}_{n.s.}$ over $h \in \mathfrak{H}_{n.s.}$ has dimension $h^0(\eta_h) - 1$. Thus if $d \geq g$, (2.2.1) holds by Riemann-Roch. On the other hand, if $d < g$, the map of $\mathfrak{w}_{n.s.}$ to $\mathfrak{H}_{n.s.}$ is generically injective. Now let $J_d \to \mathfrak{H}_{n.s.}$ be the relative Picard scheme of $C \to \mathfrak{H}_{n.s.}$ parameterizing line bundles of degree d and let $W_1 \subset J_d$ be the set of line bundles with at least one section. Then the codimension of W_1 in J_d is $\leq g - d$ and the image of $\mathfrak{w}_{n.s.}$ in $\mathfrak{H}_{n.s.}$ is the pullback of W_1 under the natural map of $\mathfrak{H}_{n.s.}$ to J_d induced by η. Thus (2.2.1) is established.

Next let $\mathfrak{w}_r$ be the inverse image of $\mathfrak{H}_r$ in $\mathfrak{w}$. We claim

(2.2.2) $$\operatorname{codim}_{\mathfrak{w}} \mathfrak{w}_r \leq r(r - g + d) .$$

Indeed, let ω' be the relative dualizing sheaf of $\pi' : C' \to \mathfrak{w}_{n.s.}$. Consider the induced map between locally free sheaves (Brill Noether matrix):

$$\Psi : \pi_*^!(\omega') \to \pi_*^!(\omega' \otimes \mathcal{O}_{\mathfrak{D}}) .$$

If Q is a closed point of $\mathfrak{w}$, then

$$h^0(C_Q^!, \mathcal{O}(\mathfrak{D})) = \dim \operatorname{coker} \Psi_Q + 1 .$$

Hence (2.2.2) follows.

Finally, we note that the fibres of $\mathfrak{w}_r$ over $\mathfrak{H}_r$ have dimension r. More algebra shows that $\operatorname{codim}_{\mathfrak{H}} W \leq \rho$, since W is a component of $\mathfrak{H}_r$ and $\mathfrak{H}_r$ is the image of $\mathfrak{w}_r$.

Suppose $W \subseteq T$. The local structure of T shows that $W \subseteq T_i$, where T_i is some component of T. Thus $\operatorname{codim}_{T_i}(W) \leq \rho - 1$. Since Φ maps S to T_i, we see

$\operatorname{codim}_S S_r \le \rho - 1$, contradicting the assumption that $\mathcal{L}$ satisfies the Brill-Noether condition. So $\mathcal{H}_r \subsetneq T$.

Finally,

$$\{h \in \mathcal{H}_r \mid \eta_h \text{ is very ample}\}$$

is open in $\mathcal{H}_r$. Hence there are points $h \in \mathcal{H}_r$ so that C_h is smooth and η_h is very ample.

§3. Let $\pi_1 : C_1 \to S_1$ be a family of smooth curves and let $Q_1,\dots,Q_p$ be sections of π_1. Assume the $Q_i(S_1)$ are disjoint.

Let ℓ be an integer, $\ell > p$.

$$S_2 = \left\{ (q_{p+1},\dots,q_\ell) \;\middle|\; \begin{array}{ll} \text{(i)} & q_i \in C_1 \\ \text{(ii)} & \pi_1(q_i) = \pi_1(q_{p+1}) \quad \text{for} \quad i \ge p+1 \\ \text{(iii)} & q_i \ne q_j \quad \text{if} \quad i \ne j \\ \text{(iv)} & q_i \ne Q_j(\pi_1(q_i)) \end{array} \right\}$$

S_2 is a locally closed subvariety of $C_1^{\ell-p}$. If we let

$$C_2 = \left\{ (q_{p+1},\dots,q_\ell,q) \;\middle|\; \begin{array}{l} (q_{p+1},\dots,q_\ell) \in S_2 \\ \pi_1(q) = \pi_1(q_i) \end{array} \right\},$$

then we get $\pi_2 : C_2 \to S_2$ and $\ell - p$ sections $Q_{p+1},\dots,Q_\ell$ where

$$Q_i(q_{p+1},\dots,q_\ell) = (q_{p+1},\dots,q_\ell,q_i) .$$

Further, we define for $i \le p$,

$$Q_i(q_{p+1},\dots,q_\ell) = (q_{p+1},\dots,q_\ell,Q_i(\pi_1(q_\ell))) .$$

Thus we have constructed a new family $\pi_2 : C_2 \to S_2$ from $\pi_1 : C_1 \to S_1$ and disjoint sections $Q_1,\dots,Q_\ell$ of π_2. We call $\pi_2 : C_2 \to S_2$ the family derived from $\pi_1 : C_1 \to S_1$ with ℓ sections.

Now let $\pi'' : C'' \to S''$ be a family of smooth curves of genus g_0 and let $Q_1,\dots,Q_p$ be sections. If s and $t \in S''$, we say s and t are associated if there is an isomorphism of C''_s to C''_t carrying $Q_i(s)$ to $Q_i(t)$. The family is good if for each $s \in S''$ there are only a finite number of t associated to s. We will assume $\pi'' : C'' \to S''$ is good. (We will be interested in the case where S'' is a point and C'' is $\mathbb{P}^1$ and Q_1, Q_2 and Q_3 are distinct points. We will also be interested when $C'' \to S''$ is a non-constant family of elliptic curves over a curve S'' and Q_1 is the zero section.) Fix ℓ. Now let $\pi' : C' \to S'$ be the family with ℓ sections $Q_1,\dots,Q_\ell$ derived from $\pi'' : C'' \to S''$.

Now let C be a curve of genus g and let L be a line bundle of degree d on C and let $\pi : \mathcal{C} \to S$ be the family with $k\ell$ sections derived for $\pi : C \to \operatorname{Spec} k$. We label these sections p_{ij} with $1 \leq i \leq k$ and $1 \leq j \leq \ell$.

Let

$$T = S \times \prod^{k} S' .$$

Let π_i denote the ith projection of T to S'. Let $C_i = T \times_{\pi_i} C'$. We denote the ℓ sections of C_i derived from $Q_1,\dots,Q_\ell$ by $Q_{i1},\dots,Q_{i\ell}$. Now given $t = (s,s_1,\dots,s_k) \in T$, we can form a stable curve $\mathcal{D}_t$ with components C and $C_{s_1},\dots,C_{s_k}$ by identifying $P_{ij}(t)$ to $Q_{ij}(t)$ to form a node. There is a family $p_1 : \mathcal{D} \to T$ and a map $C \times T \to \mathcal{D}$ so that $p_1^{-1}(t) \cong \mathcal{D}_t$ and the map $C \times \{t\} \to \mathcal{D}_t$ is the natural one.

Now fix an integer d_0. We assume $d_0 + 1 - g_0 = r$ and $d_0 > 2g_0 - 2$. Let $\mathcal{J}_t$ be the set of all line bundles $\mathfrak{m}$ on $\mathcal{D}_t$ which are isomorphic to L on C and which have degree d_0 on $C_{s_1},\dots,C_{s_k}$.

There is a variety $p : \mathcal{J} \to T$ and a line bundle $\mathfrak{m}$ on $\mathcal{J} \times_T \mathcal{D}$ so that if $\mathfrak{m}_a$ denotes the line bundle induced on $\mathcal{D}_{p(a)}$ for $a \in \mathcal{J}$, then each $\mathfrak{m}_a$ is in $\mathcal{J}_{p(a)}$, and every line bundle on $\mathcal{D}_t$ occurs uniquely as $\mathfrak{m}_a$ for some a. We leave the construction of $\mathcal{J}$ and $\mathfrak{m}$ to the reader. We note the fiber dimension of p is $k(\ell - 1) + g_0$.

We will assume $\ell > d_0$. Now let

$$A_r = \{a \in \mathcal{J} \mid h^0(\mathcal{D}_{p(a)},\mathfrak{m}_a) = r + 1\} .$$

We obtain a map

$$i_a : H^0(\mathcal{D}_{p(a)},\mathfrak{m}_a) \to H^0(C,L) ,$$

determined up to multiplication by a non-zero element of k. We note i_a is injective. Indeed, any section of $\mathfrak{m}_a$ which vanishes on C must vanish on all the nodes of $\mathcal{D}_{p(a)}$, and hence must vanish on $C_{s_1},\dots,C_{s_k}$ since $\ell > d_0$. Let G be the Grassmannian of $(r + 1)$ dimensional subspaces of $H^0(C,L)$. We have obtained a well-defined map $\Phi_1 : A_r \to G$. We leave the reader to check Φ_1 is a morphism of algebraic varieties. Further, there is a morphism $\Phi_2 : A_r \to C^{kr}$. Indeed, if $p(a) = t$,

$$\Phi_2(a) = (P_{11}(t),\dots,P_{1r}(t), P_{12}(t),\dots,P_{1r}(t),\dots)$$

Now define

$$\Psi : A_r \to G \times C^{kr}$$

by $\Psi = \Phi_1 \times \Phi_2$.

Now let $a \in A_r$ and let $p(a) = t = (s',s_1,\dots,s_k)$. If $\mathfrak{m}_a$ is very ample, we obtain an embedding $\varphi_a : \mathcal{D}_t \to \mathbb{P}^r$ defined up to projective equivalence.

DEFINITION 3.1. We say a is good if $\mathfrak{m}_a$ is very ample and further for each i there is only one curve of genus g_0 and degree d_0 passing through $\varphi_a(P_{i1}),\dots,\varphi_a(P_{i\ell})$. We also assume $\varphi_a(P_{i1}),\dots,\varphi_a(P_{i\ell})$ lie in a hyperplane H_i and are in general position in H_i.

LEMMA 3.2. <u>Suppose</u> $a \in A_r$ <u>is good. Then</u> a <u>is an isolated component of</u> $\psi^{-1}(\psi(a))$ <u>in</u> A_r.

<u>Proof</u>. The set of b in A_r so that $\mathfrak{m}_b$ is very ample is open. Suppose $\psi(a) = \psi(b)$ and $\mathfrak{m}_b$ is very ample. Then $H^0(\mathfrak{D}_{p(a)},\mathfrak{m}_a) = H^0(\mathfrak{D}_{p(b)},\mathfrak{m}_b) \subseteq H^0(C,L)$. Thus we may choose φ_a and φ_b so that $\varphi_a(\mathfrak{D}_{p(a)})$ and $\varphi_b(\mathfrak{D}_{p(b)})$ have C as a common component. Further, $\varphi_a(P_{ij}) = \varphi_b(P'_{ij})$, for $j = 1,\dots,r$, where P_{ij} and P'_{ij} are the nodes of $\mathfrak{D}_{p(a)}$ and $\mathfrak{D}_{p(b)}$ regarded as elements of C. The $\varphi_a(P_{ij})$ are in general position in a hyperplane H_i. Fix i. On the other hand, since a curve of degree d_0 and genus g_0 passes through the P'_{ij} and $d_0 > 2g_0 - 2$ and $r = d_0 + 1 - g_0$, we see the $\varphi_b(P_{ij})$ lie in a hyperplane H. But then $H = H_i$. Thus given a, the P'_{ij} are determined up to finitely many choices in $H_i \cap \varphi_a(C)$. So to establish our claim, we may assume $P_{ij} = P'_{ij}$ for all i and j. Now fixing i again, we recall there is only one curve of genus g_0 and degree d_0 passing through the $\{\varphi_a(P_{ij})\}$. Thus $\varphi_a(\mathfrak{D}_{p(a)}) = \varphi_b(\mathfrak{D}_{p_b(b)})$. Further, since $\mathfrak{m}_a = \varphi_a^*(\mathcal{O}(1))$ and $\mathfrak{m}_b = \varphi_b^*(\mathcal{O}(1))$, we see $\mathfrak{m}_a \cong \mathfrak{m}_b$. Thus if $p(a) = (s,s_1,\dots,s_k)$ and $p(b) = (s',s'_1,\dots,s'_k)$, we have $s = s'$ and s'_i is associated to s_i. Since there are only a finite number of t associated to each s_i, a is an isolated component of $\psi^{-1}(\psi(a))$.

Now let $\Delta g = (\ell - 1) + g_0$. Then the genus of $\mathfrak{D}_t$ is $g + k\Delta g$, and the degree of $\mathfrak{m}_a$ is $d + kd_0$.

PROPOSITION 3.3. <u>Suppose</u>

$$\ell + \dim S = r(\Delta g + 1) - (r + 1)d_0$$

<u>and that</u> $\mathfrak{m}_a$ <u>is good. Then</u> $\mathfrak{m}$ <u>on</u> $\mathcal{J} \times_T \mathfrak{D} \to \mathcal{J}$ <u>satisfies the Brill-Noether condition at</u> a.

<u>Proof</u>. Lemma 3.2 shows that every component of A_r passing through a has dimension at most $\dim G + kr$. On the other hand, the dimension of T is $k(\ell + \dim S)$ and the fiber dimension of $\mathcal{J} \to T$ is $k\Delta g$. So $\dim \mathcal{J} = k(\ell + \dim S + \Delta g)$. A short computation shows that the codimension of A_r in A is at least

$$(r + 1)((g + k\Delta g) - (d + kd_0) + r) .$$

So the Brill-Noether condition is satisfied.

§4. Let C be a smooth nondegenerate curve in $\mathbb{P}^r$ of degree d and genus g. We suppose $d > r + 2$, and let $L = \mathcal{O}_C(1)$. Now let $n + 1 = \dim H^0(C,L)$ and let $\varphi : C \to \mathbb{P}^n$ be the map derived from a basis of $H^0(C,L)$. Let $V \subseteq H^0(C,L)$ be the image of $H^0(\mathbb{P}^r, \mathcal{O}(1))$. We assume that there is an $s \in V$ so that if H_s is the corresponding hyperplane in $\mathbb{P}^n$, then $H_s \cap \varphi(C)$ is in general position in H_s. We further assume that for a generic $H \subseteq \mathbb{P}^r$, that $H \cap C$ does not lie on a rational normal curve (r.n.c.) in H. We note that for a generic $H \subseteq \mathbb{P}^r$, $H \cap C$ is in general position in H. We choose hyperplanes $H_1,\ldots,H_k$ and r.n.c. R_i in H_i containing $(r + 2)$ points of $H_i \cap C$.

PROPOSITION 4.1. <u>If the H_i are chosen generically, then</u>
(i) $R_i \cap R_j = \emptyset$
(ii) $\#(R_i \cap C) = r + 2$.
<u>Further, if</u> $D = C \cup R_1 \cup \cdots \cup R_k$ <u>and</u> $2k \geq h^0(C,L) - (r + 1)$, <u>we have</u> $h^0(D,\mathcal{O}_D(1)) = r + 1$.

<u>Proof</u>. Let U be the set of all hyperplanes meeting C transversally. Then for any fixed H_0, $\pi_1(U,H_0)$ operates as the full symmetric group on $H_0 \cap C$. It follows that if we set

$$\Sigma_\ell = \{(p_1,\ldots,p_\ell,H) \mid H \in U, \ p_i \in H \cap C\},$$

then Σ_ℓ is irreducible [2].

Now

$$\Gamma = \{(p_1,\ldots,p_{r+3},H) \in \Sigma_{r+3} \mid p_i \text{ lie on a r.n.c. in } H\},$$

then $\overline{\Gamma}$ is a proper closed subset of Σ by hypothesis. Since the map from Σ_{r+3} to U is quasi-finite, we can choose the H_i not in the image of $\overline{\Gamma}$. Thus $\#(R_i \cap C) = r + 2$.

Now let D be any curve of degree $\leq r$ in $\mathbb{P}^r$. We claim that for generic H_i, $R_i \cap D = \emptyset$. Indeed let

$$\Gamma = \{(p_1,\ldots,p_{r+2},H) \in \Sigma_{r+2} \mid \text{the r.n.c. through } p_1,\ldots,p_{r+2} \text{ meets } D\}.$$

We claim $\Gamma \neq \Sigma_{r+2}$. Suppose not. Pick $P_1,\ldots,P_{r+3} \in H \cap C$. Let D_i be the r.n.c. through all the P_j except P_i. Then $D_i \cap D_j \subseteq \{P_j\}$. We may assume $H \cap D \cap C = \emptyset$ and that $H \cap D$ is finite. Now D meets each D_i and $D_j \cap D_i \cap D = \emptyset$ for $i \neq j$. Hence D meets H in at least $r + 2$ points. But $\deg D \leq r + 1$. So $\Gamma \neq \Sigma_{r+2}$. Consequently, for generic H_i, $R_i \cap D = \emptyset$. Thus Proposition 4.1 i follows, since each R_i has degree $s = r - 1$.

Let $V' \supseteq V$ be any linear system of dimension $k + 1$ containing V. Choosing a basis for V' gives a morphism $\varphi_{V'} : C \to \mathbb{P}^k$. If $s \in V$, let H'_s be the hyperplane in $\mathbb{P}^k$ defined by $s = 0$. Let $V' \supseteq V'' \supseteq V$ be another linear system. If we choose s generically, the points of $H'_s \cap \varphi_{V'}(C)$ are not contained in a proper linear subspace of H'_s. Hence a monodromy argument shows that are in general position. If $(r+2)$ points of $H'' \cap \varphi_{V''}(C)$ lie in a linear subspace of dimension $r-1$, we see that $V'' = V$ or $\operatorname{codim}_{V'} V'' = 2$.

Let $D_\ell = C \cup R_1 \cup \cdots \cup R_\ell$ and let $V_\ell = H^0(D_\ell, \mathcal{O}(1))$. We claim that if the H_i are generic, then either the codimension of $V_{\ell+1}$ in V_ℓ is two or $V_{\ell+1} = V$. Indeed, let $V' = V_\ell$ and $V'' = H^0(D_{\ell+1}, \mathcal{O}(1))$. Note that the $(r+2)$ points of $H''_s \cap \varphi_{V''}(C)$ be in a linear subspace of dimension $r-1$, since a rational curve of degree $r-2$ passes through these points. Hence $V'' = V$ or $\operatorname{codim}_{V'} V'' = 2$. Proposition 4.1 is established.

Now consider the case $r = 3$. We suppose that there is an H and nine points in $H \cap C$ and a nonsingular cubic curve in H passing through the nine points. We further assume that not all the points of $H \cap C$ lie on a cubic curve. It then follows by a monodromy argument that for generic H, any nine points of $H \cap C$ lie on a smooth cubic curve. We choose the H_i generically and choose elliptic curves $E_i \subseteq H_i$ passing through nine points of C.

PROPOSITION 4.2. *If the* H_i *are chosen generically,*

(i) $E_i \cap E_j = \emptyset$

(ii) $\#(E_i \cap C) = 9$.

Further, if $D = C \cup E_1 \cup \cdots \cup E_k$ *and* $6k \geq h^0(C,L) - 4$, *we have* $h^0(D, \mathcal{O}_D(1)) = 4$.

Proof. Similar to Proposition 4.1.

§5. Let C be a curve of genus g and let L be a very ample line bundle of degree d on C with $H^1(C,L) = 0$ and let $q + 1 = h^0(C,L)$. If $V \subseteq H^0(C,L)$ is a linear system without base points with $\dim V = r+1$, we let $\varphi_V : C \to \mathbb{P}^r$ be the resulting map. We note that for $W = H^0(C,L)$, and H is a generic hyperplane, then $\varphi_W(C) \cap H$ is in general position in H.

LEMMA 5.1. *If* $h^0(C,L) \geq r+4$ *and if* V *is generic linear subspace dimension* $r+1$ *and* H *is a generic hyperplane in* $\mathbb{P}^r$, *then* $\varphi_V(C) \cap H$ *does not lie on a r.n.c. in* H. *If* $r = 3$ *and* $h^0(L) > 10$, *then* $\varphi_V(C) \cap H$ *is not contained in a cubic curve, but nine points of* $\varphi_V(C)$ *are contained in a smooth cubic curve.*

Proof. Let V' be a subspace of $H^0(C,L)$ of dimension r without base points so that $\varphi_{V'}(C)$ has only nodes and degree d. Pick $P_1,\ldots,P_{r+2}$ in $\varphi_{V'}(C)$ which lie on a r.n.c. R and pick $P_{r+3} \in \varphi_{V'}(C)$ so that $P_{r+3} \notin R$. Then the $\varphi_W(P_i)$ all lie on a hyperplane H in $\mathbb{P}^q$. Let $s \in H^0(C,L)$ be the equation of H. Let V be the subspace generated by V' and s. Then $\varphi_V(P_1),\ldots,\varphi_V(P_{r+3})$ in H are projective equivalent to $\varphi_{V'}(P_1),\ldots,\varphi_{V'}(P_{r+3})$ in $\mathbb{P}^{r-1}$.

To establish our second assertion we choose $P_1,\ldots,P_9,P_{10}$ so $\varphi_{VL}(\varphi_1),\ldots,\varphi_{V'}(P_9)$ lie on a smooth cubic E, but $\varphi_{V'}(P_{10})$ does not lie on E

and choose s to vanish on $P_1,\ldots,P_{10}$. The proof then proceeds as above.

We next turn to the proof of Proposition 1.1. We may assume $d' \leq g' + r$, since the generic line bundle with $d' \geq g' + r$ is very ample. In the notation of §3, we let $\pi_1 : C_1 \to S_1$ be $\pi_1 : \mathbb{P}^1 \to \operatorname{Spec} k$ and let Q_1, Q_2, Q_3 be three distinct points of $\mathbb{P}^1$. Let $\ell = r + 2$. So $g_0 = 0$ and let $d_0 = r - 1$. So $\Delta g = r + 1$ and $\dim S = r - 1$. Thus

$$\ell + \dim S = r(\Delta g + 1) - (r + 1)d_0 .$$

Now for a generic V of dimension $(r + 1)$ in $H^0(C,L)$, we see $\varphi_V(C) \subseteq \mathbb{P}^r$ satisfies the hypotheses of §4, Thus Proposition 4.1 constructs a good $a \in A_r$, namely $(D,\mathcal{O}(D))$, if $2k \geq h^0(C,L) - (r + 1)$. Proposition 2.2 and Proposition 3.3 gives the existence of smooth nondegenerate curves in $\mathbb{P}^r$ of degree $d + (r-1)k$ and genus $g + (r+1)k$. On the other hand, given d' and g' satisfying the hypothesis of Proposition 1.1, we can find a curve C of genus g and a line bundle L of degree d with $h^0(C,L) \geq r + 4$, $h^1(L) = 0$, and k so that $g' = g + k\Delta g$ and $d' = d + k(r-1)$, and $2k \geq h^0(C,L) - (r + 1)$. So Proposition 1.1 is established.

To establish Proposition 1.2, we let $\pi_1 : C_1 \to S_1$ be a nontrivial one dimensional family of elliptic curves and let Q_1 be the zero section. We let $\ell = 9$ and $d_0 = 3$. The proof of Proposition 1.2 proceeds as the proof of Proposition 1.1.

REFERENCES

[1] P. Deligne and D. Mumford, The irreducibility of the space of curve of given genus. Publ. IHES 36 (1969), 75-109.

[2] J. Harris, A bound on the geometric genus of projective varieties. Ann. Sci. Norm. Sup. Pis Serie IV, vo. VII, 1 (1981), 35-68.

SPRINGER FIBRES WITH NON-AMPLE NORMAL BUNDLES

Norman Goldstein
Mathematics Department
Purdue University
West Lafayette, IN 47907/USA

Let G be a complex semisimple Lie group, P a parabolic subgroup, and $Z = G/P$ a compact complex algebraic homogeneous space. I will describe a method of constructing compact submanifolds, M, of Z so that the normal bundle of M in Z, $N(M)$, is not ample i.e. there is some curve $C \subset M$ having a trivial line bundle, $\mathcal{O}_C$, contained in $N^*(M)$, the conormal space of M; see Gieseker [2 2.1] for this criterion of ampleness.

This talk was motivated by a paper of A. Papantonopoulou [4 2.8], who worked with the Grassmann varieties of linear subspaces of $\mathbb{C}^N$; see example 3, below.

We'll be considering certain subvarieties $X \subset Z$ and some dense open set of smooth points $X_0 \subset X$. Then, $N^*(X_0)$ is a locally closed subvariety of T^*Z. Put $N^*(X) = \overline{N^*(X_0)}$, the closure in T^*Z of $N^*(X_0)$. If X should happen to be smooth, then $N^*(X)$ is, of course, the conormal space of X. We will see that there is a trivial line bundle $\mathcal{O}_X \subset T^*Z|_X$, that is actually contained in $N^*(X)$.

Now, let M be any compact positive dimensional submanifold of X, and $C \subset M$ any curve. Put $\mathcal{O}_C = \mathcal{O}_X|_C$. Then

$$\mathcal{O}_C \subset N^*(X)|_C \subset N^*(M)|_C$$

i.e. $N(M)$ is not ample. This is valid even if M is contained in $X \backslash X_0$, since the pair (M, X_0) satisfies the Whitney condition A, at most points of M; see Whitney [11], Wall [10], Teissier [8].

Also, we may take M to be any compact submanifold of Z with $\dim(M) \leq \dim(X)$, and which is tangent to X_0 along some curve C_0. Let $C = \overline{C_0}$ and $\mathcal{O}_C = \mathcal{O}_X|_C$. Since $T_zM \subset T_zX_0$ for each $z \in C_0$, it is clear that $\mathcal{O}_C \subset N^*(X)|_C$, so that $N(M)$ is not ample.

We may construct such varieties, X, as follows. Let $\mathfrak{g}$ be the Lie algebra of G. For $z \in Z$, let $z^*: G \to Z$ be defined by

$$g \mapsto gz.$$

Then $z_*^{\#}: \mathfrak{g} \to T_zZ$, and this determines the well-known map

$$Z \times \mathfrak{g} \to TZ.$$

Dualizing, we obtain

$$T^*Z \to Z \times \mathfrak{g}^* \, .$$

Finally, project onto the 2nd factor.

$$\phi : T^*Z \to \mathfrak{g}^*$$

$$T^*_zZ \ni a \mapsto \phi(a) = z^{\#*}(a) \, .$$

Remark. Let $N_P \subset \mathfrak{g}$ be the nilpotent radical of P, and put $V = (G \times N_P)/P$, where P acts on $G \times N_P$ by

$$p\cdot(g,x) = (gp^{-1}, \mathrm{Ad}(p)(x)).$$

Let $\psi : V \to G$ be defined by $[g,x] \mapsto \mathrm{Ad}(g)(x)$. In [3 3.2], I describe an isomorphism

$$\begin{array}{ccc} \phi: T^*Z & \longrightarrow & \mathfrak{g}^* \\ \Big\wr & & \Big\wr \quad \mathcal{K} = \text{Killing isomorphism} \\ \psi: V & \longrightarrow & \mathfrak{g} \end{array}$$

When $P = B$, a Borel subgroup of G, the map ψ was introduced by Springer [6]. More recently, Borho-Kraft [1] have studied the case of general parabolics. In [7], Steinberg describes a map that is somewhat different, in the case $P \neq B$.

Certainly, $\psi(V)$ is contained in the nilpotent subvariety of $\mathfrak{g}$. The preceding remark shows, then, that

(1) $\mathcal{K}(\phi(T^*Z))$ consists entirely of nilpotent elements of $\mathfrak{g}$.

The group G acts on T^*Z by the codifferential of the action on Z. Let $0 \subset T^*Z$ be an orbit of this action. Let $\alpha \in \mathfrak{g}^*$, $\alpha \neq 0$, and let $Y_0 = \phi^{-1}(\alpha) \cap 0$. As is easily verified, G_α acts transitively on Y_0, so that Y_0 is smooth. Let $\pi : T^*Z \to Z$ be the natural projection, and put $X_0 = \pi(Y_0)$. Now X_0 is a dense smooth open subset of $X = \overline{X_0}$. (It is a standard fact that π embedds each fibre, $\phi^{-1}(\alpha)$, into Z.)

(2) Claim. $\phi^{-1}(\alpha)|_X \subset N^*(X)$.

(3) Remark. In this situation, $\phi^{-1}(\alpha)|_X$ spans a trivial line bundle, $\mathcal{O}_X$, in $N^*(X)$, thus completing the construction that I described earlier.

Example 1. Assume that the adjoint action of P on M_P has only finitely many orbits. (Richardson [5] shows that there is always an open orbit. However, Steinberg [7 §5 p.221] gives an example where $G = SL(8,\mathbb{C})$ and $P = B$, where there are infinitely many orbits.) Let $\alpha \in \mathfrak{g}^*$, $\alpha \neq 0$, and let Y be any irreducible component of $\phi^{-1}(\alpha)$. By finiteness, there is some orbit $0 \subset T^*Z$ for which $Y_0 = Y \cap 0$ is dense in Y. Hence, $X = \pi(Y)$ is a subvariety of Z having the property described in remark (3).

Example 2. Let Z be a smooth quadric hypersurface in $\mathbb{P}_{\mathbb{C}}$, complex projective space. Here, G is $O(n,\mathbb{C})$, the complex orthogonal group. The X's described in example 1 are the linear $\mathbb{P}^1$'s in $\mathbb{P}_{\mathbb{C}}$, that are contained in Z.

Example 3. Let Z be the Grassmannian of $\mathbb{C}^k$'s in some fixed $\mathbb{C}^N$. The X's that are described in example 1 are the sub-Grassmannians of the form $X_m = \{\mathbb{C}^k \subset \mathbb{C}^N : \mathbb{C}^m \subset \mathbb{C}^k \subset \mathbb{C}^{N-m}\}$ for some fixed subspaces $\mathbb{C}^m$ and $\mathbb{C}^{N-m}$, $m \geq 1$. Of course, each X_m is contained in some X_1, so we need only consider the X's of the form X_1.

Example 4. Let $G = SL(4,\mathbb{C})$, $P = B$ the Borel subgroup of upper-triangular matrices, and

$$\alpha = \begin{pmatrix} 0 & I_2 \\ 0 & 0 \end{pmatrix} \in \mathfrak{g},$$

considered as an element of $\mathfrak{g}^*$, via the Killing isomorphism. As in Vargas [9 2.2], one may calculate that $\phi^{-1}(\alpha)$ has two irreducible components, Y_1 and Y_2. Here, $Y_1 \simeq \mathbb{P}^1 \times \mathbb{P}^2$ and $Y_2 \simeq \mathbb{P}(\mathcal{O} + \mathcal{O}(2))$. Also, G_α has a dense orbit in each Y_i, so $X_i = \pi(Y_i)$ satisfies the conditions in remark (3).

Proof of claim (2). By continuity, we need only show that

$$\phi^{-1}(\alpha)|X_0 \subset N^*(X_0).$$

Let $a \in Y_0 = \phi^{-1}(\alpha)|X_0$ and $z = \pi(a) \in X_0$. We want to see that $a \in N_z^*(X_0)$ i.e. that $a(T_z(X_0)) = 0$. But G_α acts transitively on X_0, so that $T_z(X_0) = z_*^\#(\mathfrak{g}_\alpha)$, where $\mathfrak{g}_\alpha$ is the Lie algebra of G_α. Hence,

$$\begin{aligned} a(T_z(X_0)) = 0 &\Leftrightarrow a(z_*^\#(\mathfrak{g}_\alpha)) = 0 \\ &\Leftrightarrow z^{\#*}(a)(\mathfrak{g}_\alpha) = 0 \\ &\Leftrightarrow \alpha(\mathfrak{g}_\alpha) = 0 \end{aligned}$$

since $z^{\#*}(a) = \phi(a) = \alpha$.

View, now, α as an element of $\mathfrak{g}$, via the Killing form $(\ ,\ )$. We wish to see that $(\alpha, \mathfrak{g}_\alpha) = 0$. Let $[\ ,\ ]$ be the Lie bracket of $\mathfrak{g}$, and ad the adjoint representation of $\mathfrak{g}$ on itself. As is well-known,

$$\mathfrak{g}_\alpha = \{v \in \mathfrak{g}: [\alpha,v] = 0\} ,$$

so it remains to show that $(\alpha,v) = 0$ for each $v \in \mathfrak{g}_\alpha$. By (1), $ad(\alpha)$ is nilpotent. If $v \in \mathfrak{g}_\alpha$, then $ad(\alpha)$ and $ad(v)$ commute, so that $ad(\alpha)\circ ad(v)$ is also nilpotent. Hence, $0 = \mathrm{Trace}(ad(\alpha)\circ ad(v)) = (\alpha,v)$.

Q.E.D.

References

1. Borho, W., Kraft, H.: Über Bahnen und deren Deformationen bei linearen Aktionen reduktiver Gruppen, Comment. Math. Helvetici 54, 61-104 (1979).

2. Gieseker, D.: p-ample bundles and their Chern classes, Nagoya Math. J. 43, 91-116 (1971).

3. Goldstein, N.: Ampleness and connectedness in complex G/P, Trans. Amer. Math. Soc. 274, 361-373 (1982).

4. Papantonopoulou, A.: Curves in Grassmann varieties, Nagoya Math. J. 66, 121-137 (1977).

5. Richardson, R.: Conjugacy classes in parabolic subgroups of semi-simple algebraic groups, Bull. London Math. Soc. 6, 21-24 (1974).

6. Springer, T.: The unipotent variety of a semisimple group. Algebraic Geometry (papers presented at the Bombay Colloquium 1968), Tata Institute 1969, pp. 373-391.

7. Steinberg, R.: On the desingularization of the unipotent variety, Inv. Math. 36, 209-224 (1976).

8. Teissier, B.: Varietés polaires II. (provisional version), From courses at l'Universidad Complutense de Matrid, September 1980.

9. Vargas, J.: Fixed points under the action of unipotent elements of SL_n in the flag variety, Bol. Soc. Mat. Mexicana 24, 1-14 (1979).

10. Wall, C.: Regular stratifications. Dynamical systems, Warwick 1974. Lecture Notes in Math. 468, 332-344, Springer-Verlag, Berlin, Heidelberg, New York 1975.

11. Whitney, H.: Tangents to an analytic variety, Ann. of Math. 81, 496-549 (1965).

THE INVARIANTS OF LIAISON

C. Huneke*
Department of Mathematics
Purdue University
West Lafayette, Indiana 47907

Let X be either $\mathbb{P}_k^n$ or $\mathrm{Spec}(R)$ where R is a regular local ring. Two closed subschemes V_1 and V_2 of X are said to be directly (algebraically) linked if there is a complete intersection Z contained in $V_1 \cap V_2$ such that

$$\text{i)} \quad \underline{I}(V_2)/\underline{I}(Z) = \mathrm{Hom}_{\mathcal{O}_X}(\mathcal{O}_{V_1}, \mathcal{O}_Z)$$

and

$$\text{ii)} \quad \underline{I}(V_1)/\underline{I}(Z) = \mathrm{Hom}_{\mathcal{O}_X}(\mathcal{O}_{V_2}, \mathcal{O}_Z),$$

where $\underline{I}(Y)$ is the ideal sheaf of a closed subscheme $Y \subseteq X$.

We will write $V_1 \smile V_2$ (or $V_1 \overset{Z}{\smile} V_2$) whenever V_1 and V_2 are directly linked. If there exist $V = V_0, V_1, \ldots, V_n = V'$ such that $V = V_0 \smile V_1 \smile \ldots \smile V_{n-1} \smile V_n = V'$, then we say V and V' are <u>linked</u> and write $V \sim V'$. The linkage class of V in X, denoted $L_X(V)$, is by definition,

$$L_X(V) = \{V' \mid V \sim V'\}.$$

If $V = V_0 \smile V_1 \smile \ldots \smile V_n = V'$, then we say V is <u>evenly linked</u> to V' if n is even, and V is <u>oddly linked</u> to V' if n is odd. The corresponding linkage classes we denote by $L_X^e(V)$ and $L_X^o(V)$.

Our purpose in this note is to present a new invariant of the even and odd linkage classes of a closed subscheme V, and to apply this invariant to the study of several questions, among which is the following question: when does a linkage class $L_X(V)$ contain a complete intersection?

The definition of linkage given above is due to Peskine and Szpiro in their fundamental paper [12]. However the concept was studied by several nineteenth century mathematicians, among them Cayley [7] and M. Noether [11], and was studied in this century by Apery [1], Gaeta [8] and still later by Artin and Nagata [2].

*Supported by a NSF Postdoctoral Fellowship.

We will center this paper about two questions: what are the invariants of a linkage class, and when does a linkage class contain a complete intersection? Historically there is a motivation for both of these questions.

There are several invariants of a linkage class which are easily found. First note that in any case if V and V' are linked then V and V' are equidimensional, have no embedded components and are of the same codimension. Further in [12] it was shown that V is arithmetically Cohen-Macaulay if and only if V' is also. Thus arithmetic Cohen-Macaulayness is an invariant of a linkage class. Peskine and Szpiro go further, and prove the following proposition. (See also [2].)

Proposition 1.1. Let $X = \mathrm{Spec}(R)$, R a regular local ring. Suppose V is a closed subscheme of X which is of codimension two, equidimensional and without embedded components. Then V is Cohen-Macaulay if and only if $L_X(V)$ contains a complete intersection.

This proposition gives a complete answer to our second question in the case V has codimension two. In codimension three Cohen-Macaulayness is no longer sufficient to describe the linkage classes of complete intersections (although it is of course necessary). We will later give an example (1.15) of a arithmetically Cohen-Macaulay variety of codimension three whose linkage class does not contain a complete intersection. However, J. Watanabe [17] proved that if V is codimension three and Gorenstein, then $L_X(V)$ does contain a complete intersection. As before, this criterion fails in codimension four. In Example 1.16 we give a Gorenstein variety V of codimension four such that $L_X(V)$ does not contain a complete intersection. In fact for varieties V of codimension at least four, no simple criteria are known that guarantee $L_X(V)$ contains a complete intersection.

P. Rao completely settled the question of invariants of a linkage class for curves in $\mathbb{P}^3$. If $V \subseteq \mathbb{P}^3$ is a curve, set $M(V) = \bigoplus_{n \geq 0} H^1(\mathbb{P}^3, \underline{I}(V)(n))$. If $S = k[X_0, X_1, X_2, X_3]$ is the homogeneous coordinate ring of $\mathbb{P}^3$ then $M(V)$ is a graded S-module of finite dimension. If M_1 and M_2 are two such S-modules, say M_1 and M_2 are equivalent (written $M_1 \equiv M_2$) if M_1 and M_2 differ only up to k-duals and shifts in their grading.

Theorem 1.2. [13]: Let V and W be curves in $\mathbb{P}^3$.

i) If $V \sim W$ then $M(V) \equiv M(W)$.

ii) $M(V) = 0$ if and only if V is arithmetically Cohen-Macaulay.

iii) If M is any graded S-module of finite dimension, then $M = M(V')$ for some curve $V' \subset \mathbb{P}^3$.

The module $M(V)$ is called the Rao module of the curve V. Of course any curve in $\mathbb{P}^3$ has codimension two. Recently Rao has extended this result to the even linkage classes of codimension two varieties in $\mathbb{P}^n$. [14].

Further invariants have been discovered by Bresinsky, Schenzel, and Vogel [4] who proved that if V is Buchsbaum, so is any $V' \sim V$.

In his Paris thesis, [6], Buchweitz constructed numbers $\ell_i(V)$ which are invariants of the linkage class of V in X, and further showed how $\ell_1(V)$ relates to the deformations of V. As his construction is important for the new invariants we wish to construct, we will sketch this construction.

From now on our work is local in nature. Hence we assume for the remainder of the paper that $X = \mathrm{Spec}(S)$ where S is a regular local ring. If V is a closed subscheme of X, by $I(V)$ we denote the defining ideal of V. By m we denote the maximal ideal of S, and if $y \in G(d, m_S/m_S^2)$, the Grassmanian of d-planes in m_S/m_S^2, we will identify y with a partial system of parameters $y_1, \ldots, y_d$ in S. (To do this fix a basis of m_S.)

If V is a Cohen-Macaulay closed subscheme of X, we denote the dualizing sheaf of V by ω_V. If F is a Cohen-Macaulay coherent module of O_X of dimension d, there exists an open set $U_F \subseteq G(d, m_S/m_S^2)$ such that if $y \in U_F$, then $y_1, \ldots, y_d$ are a system of parameters of $H^0(X,F)$. If $y \in U_F$, let $\overline{X} = \mathrm{Spec}(S/y_1, \ldots, y_d))$, and put $\overline{V} = V \times \overline{X}$. We define

$$\ell_i(V) = \min_{y \in U_V} \lg(\mathrm{Tor}_i(O_{\overline{V}}, \omega_{\overline{V}})) - \binom{k}{i} e(V),$$

where $e(V)$ is the multiplicity of V and $\lg(\ )$ denotes length. The minimum is obtained on an open subset of U_V. Buchweitz is able to show,

Theorem 1.3. [6]:

1) If $V \sim W$ then $\ell_i(V) = \ell_i(W)$ for all i.
2) $\ell_i(V) = 0$ for $0 \leq i \leq m$ if and only if $\mathrm{Tor}_i(O_V, \omega_V)$ are Cohen-Macaulay for $0 \leq i \leq m$.
3) If V is a complete intersection, then $\ell_i(V) = 0$ for all i.

Thus if $L_X(V)$ contains a complete intersection, $\ell_i(V) = 0$ for all i, and $\mathrm{Tor}_i(O_V, \omega_V)$ is Cohen-Macaulay for all i. In particular V is strongly non-obstructed. [9].

Now fix a regular local ring S and let $R = S/I$. Fix a generating set $z_1, \ldots, z_n$ of I. The Koszul complex of the z_i is the complex

$$\bigotimes_{i=1}^{n} 0 \to S \xrightarrow{z_i} S \to 0,$$

and we denote this complex by $K_\bullet(z;S)$. (Here the complexes $0 \to S \xrightarrow{z_i} S \to 0$ are concentrated in degrees 1 and 0.)

If M is any S-module, by $H_\bullet(\underline{z};M)$ we denote the homology

$$H_\bullet(K_\bullet(\underline{z};s) \otimes M).$$

The module $H_\bullet(\underline{z};M)$ is called the Koszul homology of I with respect to M and is annihilated by $I + \mathrm{ann}(M)$. If $d = \dim R$, there is an open subset $U_R \subseteq G(d, m_S/m_S^2)$ such that if $y \in U_R$, then $y_1,\dots,y_d$ form a system of parameters in R. By "$\bar{\ }$" denote the map from S to $S/(y_1,\dots,y_d)$.

Definition 1.5. Fix the notation as above. Then we define $k_i(\underline{z};S)$ to be

$$\min_{y \in U_R} \ \lg(H_i(\underline{z};\bar{S})) - \binom{n-k}{i} e(R),$$

where $\underline{z} = z_1,\dots,z_n$, $k = \operatorname{codim} R$, and $e(R)$ is the multiplicity of R.

Several explanatory comments can be made.

1) Since $\operatorname{depth}_I \bar{S} = k$, $H_i(\underline{z};\bar{S}) = 0$ if $i > n-k$. If $i = n-k$, then $H_{n-k}(\underline{z};\bar{S}) \simeq K_{\bar{R}}$, the canonical module of $\bar{R}$.
2) $\lg H_i(\underline{z};\bar{S}) < \infty$ since both I and $y_1,\dots,y_d$ annihilate these modules.
3) If R is a domain, then
$$\binom{n-k}{i} e(R) = \operatorname{rank} H_i(\underline{z};S)\cdot e(R) = e(H_i(\underline{z};S)).$$
4) The integers $k_i(\underline{z};S)$ depend upon the generating set of I.
5) If $\dim R = d$, then
$$\dim H_i(\underline{z};S) = d$$
for any generating set $\underline{z}$ of I.

Set $f(\underline{z};S) = \sum_{i=0}^{n-k} k_i(\underline{z};S)t^i$, a polynomial in $\mathbb{Z}[t]$.

Definition 1.6. By $P_R^S(t)$ we denote the polynomial $f(\underline{z};S)/(1+t)^m$ where $(1+t)^m$ divides $f(\underline{z};S)$ but $(1+t)^{m+1}$ does not.

Our main result is,

Theorem 1.7. Let the notation be as above. Then the following statements hold.

i) $P_R^S(t)$ does not depend upon the generating set of $I = I(R)$.

ii) If $R' \in L_S^e(R)$, then $P_{R'}^S(t) = P_R^S(t)$.

iii) If R is reduced, $P_R^S(t)$ is divisible by t^i if and only if $H_j(\underline{z};S)$ is a Cohen-Macaulay module for all $j < i$. In particular, t divides $P_R^S(t)$ if and only if R is Cohen-Macaulay while $P_R^S(t) = 0$ if and only if $H_j(\underline{z};S)$ is Cohen-Macaulay for all $j \geq 0$.

iv) Write $P_R^S(t) = t^i P_1(t)$ where t does not divide $P_1(t)$. Then

$$P_1(\tfrac{1}{t})t^{\deg P_1} = P_1(t).$$

v) If R is a complete intersection, then $P_R^S(t) = 0$.

The assertions i) and v) are trivial, while iii) follows from the following by standard techniques from the theory of multiplicities in Serre [15]. If M is an S-module and $y_1,\ldots,y_d$ is an S-sequence such that $\lg(M \otimes \overline{S})$ is finite $(\overline{S} = S/(y_1,\ldots,y_d))$ then the intersection multiplicity of M and $\overline{S}$ is defined to be

$$\chi(M_1,\overline{S}) = \sum_{i=0}^{\infty} (-1)^i \lg(\mathrm{Tor}_i(M_1,\overline{S})). \tag{1}$$

The sum on the right is finite since $\mathrm{pd}\overline{S}$ is finite. Serre proves the following facts:

1) $\chi(M,\overline{S}) = 0$ if $\dim M + \dim \overline{S} < \dim S$.

2) $\chi(M,\overline{S}) > 0$ if $\dim M + \dim \overline{S} = \dim S$.

3) $\lg(M \otimes \overline{S}) - \chi(M,\overline{S}) \geq 0$ and is equal to zero if and only if $\mathrm{Tor}_i(M,\overline{S}) = 0$ for $i \geq 1$, that is if and only if M is a Cohen-Macaulay module.

If R is reduced, then

$$\chi(H_i(\underline{z};S),\overline{S}) = \binom{n-k}{i} e(R).$$

Thus,

$$k_i(\underline{z};S) = \lg H_i(\underline{z};\overline{S}) - \chi(H_i(\underline{z};S),\overline{S})$$

for some $y \in U_R$. The assertion of iii) would follow immediately from 3) above if $H_i(\underline{z};S) \otimes \overline{S} = H_i(\underline{z};\overline{S})$. In general this equality does not hold. If $H_j(\underline{z};S)$ is Cohen-Macaulay for $j < i$, however, the equality does hold and iii) follows from an easy induction.

The main assertion is of course ii), which we do not prove here. Rather we discuss several corollaries which follow from Theorem 1.7.

Firstly, Theorem 1.7 allows us to define another polynomial $Q_R^S(t)$ as follows: Choose any R' such that $R' \smile R$. (<u>Directly linked</u>.) Set $Q_R^S(t) = P_{R'}^S(t)$. If $R'' \smile R$ is another ring directly linked to R

then $P^S_{R'}(t) = P^S_{R''}(t)$ since R' and R'' are evenly linked. Thus $Q^S_R(t)$ does not depend on the choice of R'. We refer to $Q^S_R(t)$ as the <u>odd polynomial</u> of R and $P^S_R(t)$ as the <u>even polynomial</u> of R.

<u>Proposition 1.8</u>. Suppose R is Gorenstein. Then $Q^S_R(t) = 0$.

<u>Proof</u>: Let R' R and write $R' = S/J$, $R = S/I$, and suppose the linkage is given by the complete intersection $S/\underline{x}$. Then $R = K_R = \mathrm{Hom}(R, S/\underline{x}) = J/\underline{x}$. Therefore either J is a complete intersection or is generated by codim $R' + 1$ elements. If we choose $z_1, \ldots, z_p$ to be a minimal generating set of J, then if J is a complete intersection,

$$H_i(\underline{z};\bar{S}) = 0 \quad \text{for} \quad i \geq 1$$

if $y \in U_{R'}$. In particular $P^S_{R'}(t) = 0$ in this case (Theorem 1.7 v)).

In the other case, $H_i(\underline{z};\bar{S}) = 0$ for $i > 1$ and $H_1(\underline{z};\bar{S}) = K_{\bar{R}}$. We have, $k_0 = \lg(\bar{R}') - e(R')$ and $k_1 = \lg(K_{\bar{R}'}) - e(R')$.

Since R is Gorenstein, R' is Cohen-Macaulay and so $k_0 = 0$. By duality, $\lg(K_{\bar{R}'}) = \lg(\bar{R}')$ and so $k_1 = 0$ also. Thus $Q^S_R(t) = P^S_{R'}(t) = 0$.

<u>Corollary 1.9</u>. [10] Suppose $L_S(R)$ contains a complete intersection. Then $P^S_R(t) = 0$. If R is reduced, then $H_i(\underline{z};S)$ are Cohen-Macaulay for all $i \geq 0$, where $\underline{z} = (z_1, \ldots, z_n) = I(R)$.

<u>Proof</u>: Write $R \smile R_1 \smile R_2 \smile \ldots \smile R_n$ where R_n is a complete intersection. By Theorem 1.7 v), $P^S_{R_n}(t) = 0$. Hence if n is even, $P^S_R(t) = 0$ by Theorem 1.7 ii). If n is not even, $n - 1$ is even and so $P^S_R(t) = P^S_{R_{n-1}}(t)$. However, R_n is Gorenstein. From Proposition 1.8, $P^S_{R_{n-1}}(t) = Q^S_{R_n}(t) = 0$. In either case $P^S_R(t) = 0$!

The last statement follows immediately from Theorem 1.7, iii).

Corollary 1.9 gives a useful necessary condition to determine if $L_S(R)$ can contain a complete intersection. To see this, we consider the module $T_2^{R/S}$. This is defined by the exact sequence

$$0 \to T_2^{R/S} \to H_1(\underline{z};S) \to R^n \to I/I^2 \to 0. \tag{2}$$

Here $(z_1, \ldots, z_n) = I = I(R)$ and the maps are the obvious ones.

It is easily observed from (2) that if R is reduced, and $H_1(\underline{z};S)$ is Cohen-Macaulay, then $T_2^{R/S} = 0$. We therefore obtain the corollary,

<u>Corollary 1.10</u>. Suppose R is reduced and $L_S(R)$ contains a complete intersection. Then $T_2^{R/S} = 0$.

The module $T_2^{R/S}$ has an alternate description. (For example see [16].) Namely $T_2^{R/S} = \ker(\mathrm{Sym}_2 I \to I^2)$. Here $\mathrm{Sym}_2 I$ = the second symmetric power of I. This characterization allows us to give a concrete description of $T_2^{R/S}$. Adjoin n new variables $T_1,\ldots,T_n$ to S and let q equal the ideal of $S[T_1,\ldots,T_n]$ generated by the set of all $\sum_{i=1}^{n} a_i T_i$ where $a_i \in S$ and $\sum_{i=1}^{n} a_i z_i = 0$. Let q' = ideal generated by all forms $F(T_1,\ldots,T_n)$ in $S[T_1,\ldots,T_n] = A$ such that $F(z_1,\ldots,z_n) = 0$. Clearly $q' \supset q$. Then $T_2^{R/S} = q'[2]/q[2]$, where by $I[n]$ we denote the forms of degree n in I.

If $z_1,\ldots,z_n$ is a minimal generating set of I, then $q[2]$ is clearly contained in $m_S A[2]$. Thus if there is a polynomial $F \in q'[2]$ with unit coefficients, $T_2^{R/S} \neq 0$.

We apply these observations to determinantal varieties. Let V be a vector space over k of dimension r and W a vector space over K of dimension s $(r \leq s)$. We define $D_{r,s}^t = \{a \in \mathrm{Hom}_K(V,W) \mid \text{rank } a \leq t\}$ $(1 \leq t \leq r)$. If we embed $D_{r,s}^t \subseteq \mathrm{Hom}(V,W)$, then $I(D_{r,s}^t)$ is defined by the ideal $I_{t+1}(X)$ generated by the $t+1$ size minors of the generic r by s matrix $X = (x_{ij})$. Set $S = k[[x_{ij}]]$, $I = I_t(X)S$, $R = S/I$.

<u>Corollary 1.11</u>. If either $1 < t < r < s$ or $t = r < s - 1$, then $L_S(R)$ does not contain a complete intersection.

<u>Proof</u>: If $1 < t < r < s$ or $t = r < s - 1$ there are non-zero Plucker relations on the generators of $I_t(X)$. These relations are quadratic polynomials with unit coefficients. The comments above show $T_2^{R/S} \neq 0$, and Corollary 1.10 gives the needed conclusion.

<u>Corollary 1.12</u>. Suppose either i) $\mathrm{codim}_S R = 2$ and R is Cohen-Macaulay or ii) $\mathrm{codim}_S R = 3$ and R is Gorenstein.
Then $H_i(\underline{z};S)$ is Cohen-Macaulay for all i, where $\underline{z} = I = I(R)$.

<u>Proof</u>: For i) see also [3]. In either case i) or ii) there is a ring of the form $T = S[[T_1,\ldots,T_n]]$ and a domain $B = T/J$ such that in case i) B is Cohen-Macaulay of codimension two, in case ii) B is Gorenstein of codimension three (see [B-E]), and B specializes to R. That is, there is a regular sequence $a_1,\ldots,a_m$ on T ($a_i = T_i - b_i$ for some $b_i \in S$) such that

$$S = \overline{T} = T/(a_1,\ldots,a_m)$$

and

$$R = B \otimes_T \overline{T}.$$

By the remarks in the introduction we know that $P_B^T(t) = 0$. Since B is a domain we conclude that $H_i(J;B)$ is Cohen-Macaulay for all i. (Choose any generating set of J.) In this case,

$$H_i(J;B) \otimes_T \overline{T} = H_i(I;R)$$

is also Cohen-Macaulay.

Now assume $\operatorname{codim} R = \dim S$ and R is Gorenstein. We wish to interpret $k_1(\underline{z};S)$ where $\underline{z} = (z_1,\ldots,z_n)$ is a minimal generating set of $I = I(R)$. There is an exact sequence (see above),

$$0 \to T_2^{R/S} \to H_1(\underline{z};S) \to R^n \to I/I^2 \to 0.$$

Each term in this exact sequence has finite length. Therefore,

$$\lg(I/I^2) + \lg(H_1(\underline{z};S)) = n \lg(R) + \lg(T_2^{R/S}).$$

Since R is Gorenstein, $\lg(T_2^{R/S}) = \lg(T^2_{R/S})$, while

$$\lg(I/I^2) - k \lg(R) = \ell_1(R;S),$$

and

$$\lg(H_1(\underline{z};S)) - (n-k)\lg(R) = k_1(R;S).$$

Therefore,

$$k_1(\underline{z};S) + \ell_1(R;S) = \lg(T^2_{R/S}). \tag{3}$$

<u>Corollary 1.13</u>. Suppose R is Gorenstein and $\dim R = 0$. Then

$$k_1(\underline{z};S) = \ell(T^2_{R/S}) - \ell(T^1_{R/S}) + \ell(T^0_{R/S}).$$

If $L_S(R)$ contains a complete intersection then $T^2_{R/S} = 0$.

<u>Proof</u>: By [6], $\ell_1(R;S) = \ell(T^1_{R/S}) - \ell(T^0_{R/S})$. The first assertion of the corollary follows immediately from (3). If $L_S(R)$ contains a complete intersection then both $\ell_1(R;S) = 0$ (Theorem 1.3) and $k_1(\underline{z};S) = 0$ (Theorem 1.7). It follows that $T^2_{R/S} = 0$.

We close with some examples and questions.

<u>Example 1.14</u>. Suppose $\operatorname{codim} R = 2$. Then $P_R^S(t) = \lg(\overline{R}) - e(R)$, is a constant.

<u>Example 1.15</u>. If $S = k[[x_{ij}]]$, $X = (x_{ij})$ is a 2 by 4 matrix and $I = I_2(X)$, $R = S/I$ then $P_R^S(t) = 2t$. R is a Cohen-Macaulay ring of codimension three which is not in the linkage class of a complete intersection as $P_R^S(t) \neq 0$. One can show $P_R^S(t) = Q_R^S(t)$. This example and the others below were computed with the help of the computer program of M. Stillman.

Example 1.16. If $S = k[[X_{ij}]]$, $X = (X_{ij})$ a 3×3 matrix and $R = S/I_2(X)$, then $P_R^S(t) = 5t$. R is a Gorenstein ring of codimension four which is not in the linkage class of a complete intersection, as $P_R^S(t) \neq 0$. However $Q_R^S(t) = 0$ by Proposition 1.8. This shows that in general P_R^S and Q_R^S are not equal.

Example 1.17. Let $R = k[xs,ys,zs,xt,yt,zt]$ where $x^3 + y^3 + z^3 = 0$. R is a 3-dimensional normal non-Cohen-Macaulay ring of embedding dimension six. $P_R^S(t) = t^2 + 6t + 1$.

Example 1.18. Let $I = (X_1,X_2,X_3) \cap (Y_1,Y_2,Y_3)$, $S = k[X_1,X_2,X_3,Y_1,Y_2,Y_3]$, and $R = S/I$. Then $P_R^S(t) = t^2 + 3t + 1$.

Example 1.19. The polynomial equation of Theorem 1.7 v) follows from the fact that if the deviation of R is d, then

$$k_i = k_{d-i}.$$

In addition one can show

$$\sum_{i=0}^{d} (-1)^i k_i = 0.$$

From these observations one can prove that if t^2 divides $P_R^S(t)$, and $d \leq 4$, then $P_R^S(t) = 0$.

We close by asking several questions.

1. Is there a good bound on $\deg P_R^S(t)$? For instance is $\deg P_R^S(t) \leq \operatorname{codim}_S R$?

2. Does there exist a variety $V \subseteq X = \operatorname{Spec}(S)$ such that $P_V^X(t) = Q_V^X(t) = 0$ but with $L_X(V)$ not containing a complete intersection?

3. Is there an interpretation of the k_i in terms of the higher T^j?

4. What arithmetic properties of R are represented by the roots of $P_R^S(t)$?

REFERENCES

1. R. Apery, Sur les Courbes de premiere espèce de l'espace à trois dimensions, C. R. Acad. Sci. Paris Ser. A-B 220(1945), 271-272.

2. M. Artin and M. Nagata, Residual intersections in Cohen-Macaulay rings, J. Math. Kyoto Univ. 12(1972), 307-323.

3. L. Avramov and J. Herzog, The Koszul algebra of a codimension two embedding, to appear, J. of Alg.

4. H. Bresinsky, P. Schenzel, and W. Vogel, On liaison, arithmetical Buchsbaum curves and monomial curves in $\mathbb{P}^3$, Queen's University preprint, (1981).

5. D. Buchsbaum and D. Eisenbud, Algebra structures for finite free resolutions and some structure theorems for ideals of codimension three, Amer. J. Math. 99(1977), 447-485.

6. R.-O. Buchweitz, Contributions a la theorie des singularities, thesis, Paris (1981).

7. A. Cayley, Note sur les hyperdéterminants, J. Reine Angew. Math. 34(1847), 148-152.

8. F. Gaeta, Quelques progrès récents dans la classification des variétiès algèbriques d'un espace projectif, Deuxièume Collogue de Géométrie Algèbrique Liège, C. B. R. M., 1952.

9. J. Herzog, Deformation von Cohen-Macaulay algebren, J.f.d.r.u.a. Math. 318(1980), 83-105.

10. C. Huneke, Linkage and the Koszul homology of ideals, Amer. J. Math., 104(1982), 1043-1062.

11. M. Noether, Zur Grundlegung der Theorie der algebraischen Raumkurven. Abh. der Konigl. Preuss. Akad. Wiss. zu Berlin, Verlag der Koniglichen Akademie der Wissenschaften, Berlin 1883.

12. C. Peskine and L. Szpiro, Liaison des varietés algébriques, Inventiones Math. 26(1974), 271-302.

13. P. Rao, Liaison among curves in $\mathbb{P}^3$, Invent. Math. 50(1979), 205-217.

14. P. Rao, Liaison equivalence classes, Math. Annalen, 258(1981), 169-173.

15. J. P. Serre, Algebre Locale; Multiplicities, Springer Lecture Notes 11, Springer-Verlag, Berlin-Heidelberg-New York, 1965.

16. A. Simis and W. Vasconcelos, The syzygies of the conormal module, Amer. J. Math. 103(1981), 203-224.

17. J. Watanabe, A note on Gorenstein rings of embedding codimension 3, Nagoya Math. J. 50(1973), 227-232.

On étale coverings of the affine space

by

T. Kambayashi (DeKalb) and V. Srinivas (Princeton)

In the affine (or noncomplete) algebraic geometry the varieties equipped with certain types of morphisms to or from an affine space play a significant role. At the Ann Arbor conference of November 1981 one of the present authors gave a talk on such varieties entitled "Varieties over and under the affine space", discussing results of M. Miyanishi, R. V. Gurjar and ourselves. In the report at hand we treat one-half of that talk, confining ourselves to varieties étale and finite over an affine space.

In this paper whenever a ground field, denoted by k, enters in the picture it is understood to be <u>algebraically closed</u>. Even when no ground fields are involved our schemes are always normal and integral. By $\mathbb{P}^n$ and $\mathbb{A}^n$ we denote respectively the n-dimensional projective and affine spaces over k.

Let $Y \to \mathbb{A}^n$ be an étale cover (or covering) of $\mathbb{A}^n$. By this is meant the map is both étale <u>and</u> finite. The main question then is, "What does Y look like?" It is a strictly characteristic p question, as otherwise the answer is trivial. One could rephrase the question in terms of the algebraic fundamental group $\pi_1(\mathbb{A}^n)$ (see §1 below). However put and as far as we know, the question has not been answered very successfully by anyone, even at $n = 1$ level. (Thus, rather embarrassingly, one doesn't know much about separable algebras over a one-variable polynomial ring $k[X]$!) We offer what little we have discovered in §2 below.

1. <u>Étale covers of</u> $\mathbb{A}^n$ <u>and</u> $\mathbb{P}^n$.

Let S be a normal variety over k, i.e., normal integral scheme of finite type over the ground field k, and $K := k(S)$ the rational function field of S. Fixing an algebraic closure $\overline{K}$ of K, let us define <u>the algebraic fundamental group</u> $\pi_1(S)$ and the <u>tame fundamental group</u> $\pi_1^t(S)$ as follows: For any finite extension field L, $K \subseteq L \subseteq \overline{K}$, denote by S_L the normalization of S in L, which, then, is endowed with a natural finite map $S_L \to S$. Let Σ be the set of all those <u>finite Galois</u> extensions L of K for which the map $S_L \to S$ is étale. Further, if $p = \mathrm{char}(k) > 0$, let Σ^t be the subset of Σ consisting of those $L \in \Sigma$ whose degree $[L : K]$ over K be <u>not</u> divisible by p. It is clear that Σ and Σ^t each form an inductive system relative to the inclusion relation $L \hookrightarrow L'$ of fields. Now define

$$\pi_1(S) := \varprojlim\{\mathrm{Gal}(L/K) : L \in \Sigma\},$$
$$\pi_1^t(S) := \varprojlim\{\mathrm{Gal}(L/K) : L \in \Sigma^t\}.$$

(Here, $\mathrm{Gal}(L/K)$ stands for the Galois group of L over K.) The following facts are rather well-known:

1.1. THEOREM. $\pi_1(\mathbb{P}^n) = \{1\}$ and $\pi_1^t(\mathbb{A}^n) = \{1\}$.

We sketch relevant arguments. Let $S = \mathbb{A}^n$ or $\mathbb{P}^n$, and let $f: Y \to S$ be an <u>étale Galois cover</u>, i.e., an étale finite morphism with Y a normal variety over k and with function fields $k(Y)/k(S)$ forming a Galois extension. In case $p = \mathrm{char}(k) > 0$ and $S = \mathbb{A}^n$, assume additionally that $p \nmid \deg(f)$. For a general hyperplane $H \subset S$ the inverse image $f^{-1}(H)$ is irreducible by virtue of Bertini's First Theorem. Replace S by H, Y by $f^{-1}(H)$ and f by its restriction to $f^{-1}(H)$, and we get an étale Galois cover of H of the same degree as before. Repeat this replacement procedure as long as the dimensions are greater than one. So, we come to an étale Galois cover $f_1: Y_1 \to S_1$ with $S_1 = \mathbb{A}^1$ or $\mathbb{P}^1$, and $p \nmid \deg(f_1) = \deg(f)$ if $S_1 = \mathbb{A}^1$, $p > 0$. At this point we substitute for S_1 and Y_1 their complete normal models if they are not already complete; then we have a finite Galois map $f_1: Y_1 \to \mathbb{P}^1$ where Y_1 is a nonsingular complete curve and f_1 is unramified above all points of $\mathbb{P}^1$ except possibly at infinity. Moreover, above the point P_∞ at infinity, f_1 is tamely ramified at worst. Indeed, if $Q_1,\dots,Q_r$ are the places of $k(Y_1)$ lying over P_∞ with ramification degree $e \geq 1$, then $er = d := \deg(f_1)$; since $p \nmid d$ by assumption, $p \nmid e$. Consequently, Hurwitz' Formula applies to this situation: if g denotes the genus of Y_1,

$$2g - 2 = -2d + r(e-1) = -d - r,$$

whence $2g = -d + (2-r) \leq 1 - d$, forcing the conclusion that $d = 1$, $g = 0$.

1.2. The first part of 1.1 says there are no étale Galois covers of $\mathbb{P}^n$. Since every étale cover is enlarged to an étale Galois cover (see 2.4 below), one can conclude that $\mathbb{P}^n$ has no étale covers. The second part of 1.1 tells us likewise that, in characteristic zero, there are no étale covers of $\mathbb{A}^n$. When $p = \mathrm{char}(k) > 0$, however, one can only say that there exists no étale <u>Galois</u> cover of $\mathbb{A}^n$ whose <u>degree is non-divisible by</u> p. Note that if either of these underscored conditions is dropped there arise many étale covers of $\mathbb{A}^n$. Indeed:

(a) Omit Galoisian requirement and one gets <u>Abhyankar's Example</u> [1; Th. 1, p. 830]: Let

$$\begin{cases} R := k[t], \text{ a one-variable polynomial ring, and} \\ A := R[X]/(X^{p+1}-tX+1). \end{cases}$$

Then it is easily verified that $\operatorname{spec} A \to \operatorname{spec} R = \mathbb{A}^n$ is an étale cover of degree $p + 1$. There are two places at infinity of spec A with ramification indices p and 1 each over the point at infinity of $\mathbb{A}^1$. (This example can be generalized to yield many other such coverings.)

(b) Artin-Schreier extensions. If the non-divisibility by p is omitted, one obtains these classical examples: $R = k[t]$ as before, and let

$$B := R[X]/(X^p - X - f(t))$$

for any $f(t) \in k[t]$ provided $f(t)$ is not expressible as $g^p - g$ by some $g \in k[t]$. This is an étale Galois cover of degree p over $\mathbb{A}^1$.

2. Étale covers of $\mathbb{A}^n$ of degree p.

As a first step toward understanding $\pi_1(\mathbb{A}^n)$ in case $p = \operatorname{char}(k) > 0$, we examine étale covers of $\mathbb{A}^n$ of degree p. If such a cover is Galoisian, then it is always of Artin-Schreier type. In fact, quite generally, we have

2.1. THEOREM. Every étale Galois cover of degree p of an affine scheme $S = \operatorname{spec} R$ in characteristic p is of Artin-Schreier type.

In fact, if Y is such a cover, then Y is a $(\mathbb{Z}/p\mathbb{Z})$-torsor in étale topology over $S = \operatorname{spec} R$. Let F denote the Frobenius morphism. Since

$$0 \to \mathbb{Z}/p\mathbb{Z} \to \mathbb{G}_a \to \mathbb{G}_a \to 0$$

is exact as étale sheaves over S, there is an exact sequence

$$H^\circ(S,\mathbb{G}_a) \xrightarrow{F-1} H^\circ(S,\mathbb{G}_a) \longrightarrow H^1(S,\mathbb{Z}/p\mathbb{Z}) \longrightarrow H^1(S,\mathbb{G}_a)$$

where S is considered as an étale site. But the sheaf $\mathbb{G}_a$ over S arises from an obvious quasi-coherent sheaf of $\mathcal{O}_S$-modules, so these étale cohomology groups except the third one agree with Zariski cohomology groups and give

$$H^\circ(S,\mathcal{O}_S) \xrightarrow{F-1} H^\circ(S,\mathcal{O}_S) \xrightarrow{d} H^1(S,\mathbb{Z}/p\mathbb{Z}) \to H^1(S,\mathcal{O}_S) = 0.$$

Therefore, $H^1(S,\mathbb{Z}/p\mathbb{Z})$ is the cokernel of $F - 1 : R \to R$ (sending $r \in R$ to $r^p - r$), and the construction of the connecting map d in the standard way shows each element of $H^1(S,\mathbb{Z}/p\mathbb{Z})$ is represented as $Y = \operatorname{spec} B$ by a ring extension

$$B = R[X]/(X^p - X - a)$$

for some $a \in R$. (B is trivial iff $a = r^p - r$ for some $r \in R$.) For more details, the reader might consult Milne's book [3; III, §4].

2.2. Remark. The preceding theorem is proved by direct computations under certain additional assumptions on R (Noetherian, factorial, etc.) in Miyanishi's recent paper [4; Th. 1.1].

We now turn to arbitrary covers of $\mathbb{A}^n$ of degree p, and ask ourselves of a

2.3. Question. Is every étale cover of $\mathbb{A}^n$ of degree p Galoisian (and, hence, of Artin-Schreier type in view of 2.1)?

We are unable to answer this question except when $p = 2$ or 3, although in all likelihood the answer in general is 'yes'. In order to connect an étale cover with one that is Galoisian it is convenient to introduce the Galois closure of an étale morphism. This is nothing new, especially in the finite étale case, as treated by Serre [5; §1.5], Abhyankar [1; §5, pp. 843-845], and Villamayor [6], among many others. But the following very simple treatment in the affine normal case may be worth recording:

2.4 LEMMA. Let $f: Y \to X$ be a dominant, finite-type étale morphism of Noetherian normal integral affine schemes Y and X, with their respective function fields L and K. Then, the smallest Galois extension field $\tilde{L}$ over K containing L has an affine model $\tilde{Y}$ equipped with an étale morphism $\tilde{f}: \tilde{Y} \to X$ which factors through f.

Proof. Let Y = spec B, X = spec A, so that B is a finitely-generated A-algebra. Let Γ be the set of all K-isomorphisms of L into $\tilde{L}$, and let $\tilde{B}$ be the A-algebra generated inside $\tilde{L}$ by the σ-transforms B^σ for all $\sigma \in \Gamma$. Let $\tilde{Y}$:= spec $\tilde{B}$, and $\tilde{f}: \tilde{Y} \to Y \to X$ the obvious map. Then $\tilde{f}$ is of finite type, clearly; $\tilde{f}$ is unramified since $\Omega^1(\tilde{B}|A) = 0$. In fact, $\Omega^1(\tilde{B}|A)$ is generated as $\tilde{B}$-module by $\{d_{\tilde{B}|A}(b^\sigma): b \in B, \sigma \in \Gamma\}$ but, for each b and σ, $d_{\tilde{B}|A}(b^\sigma) = 0$ because $d_{B^\sigma|A}(b^\sigma) = 0$. The flatness of $\tilde{f}$ results from its unramifiedness and the normality of X by virture of (EGA IV, 18.10.1).

Q.E.D.

If f is in addition finite in 2.4 above, then the ring $\tilde{B}$ agrees with the integral closure of A in $\tilde{L}$. Important note: In contrast to ones in the étale-finite topology, a covering in étale (non-finite) topology does not generally admit a Galois closure, since the $\tilde{f}$ above fails to inherit the surjectivity of f. A case in point is the following

2.5. *Example*. Let $X = \mathbb{A}^1 = \operatorname{spec} k[s]$ with $p = \operatorname{char}(k) > 0$, and $Y = \operatorname{spec}(k[s,T]/(sT^p-T-s))$. Then, the natural map $f\colon Y \to X$ is étale, surjective but not finite. In the notation of the proof of 2.4, Γ is provided by the $p-1$ mappings that send t $(:=T \bmod (sT^p-T-s))$ to $t + s^{1/(1-p)}$, plus the neutrál mapping. Consequently, $s^{1/(1-p)} \in \tilde{B}$, hence $s^{-1} \in \tilde{B}$. Thus, s is a prime element in B and a unit in $\tilde{B}$.

We do not know if this is a strictly characteristic p pathological phenomenon

We now return to étale covers of $\mathbb{A}^n$ of degree $p > 0$ (see 2.3).

2.6. THEOREM. *Let S be a Noetherian normal integral affine scheme. Then, every étale cover of S of degree 2 is Galoisian; if S is furthermore a characteristic-two scheme, then every such cover is of Artin-Schreier type.*

This is quite clear in view of 2.4 and 2.1.

2.7. THEOREM. *Let S be a Noetherian normal integral affine scheme, and assume that no nontrivial étale Galois cover of S of degree 2 exists* (e.g. $S = \mathbb{A}^n_k$, $\operatorname{char}(k) \neq 2$). *Then, every étale cover of S of degree 3 is Galoisian. In case S is of characteristic 3, such a cover is of Artin-Schreier type.*

Proof. Let $f\colon Y \to S$ be an étale cover of degree 3, and construct the Galois closure $\tilde{f}\colon \tilde{Y} \to Y \to S$ as done in 2.4. Since f is finite, $\tilde{f}$ is an étale Galois cover with Galois group Γ a subgroup of S_3 (symmetric group on 3 letters). The group Γ contains a subgroup of index 3 corresponding to Y, so $\Gamma \simeq S_3$ or $\mathbb{Z}/3\mathbb{Z}$. If $\Gamma \simeq S_3$ then Γ would contain a subgroup $\Delta \simeq A_3$ (alternating group on 3 letters); in that case the quotient scheme $\tilde{Y}/\Delta$ would be an étale Galois cover of S of degree 2, an impossiblity under our hypothesis. Consequently, $\Gamma \simeq \mathbb{Z}/3\mathbb{Z}$ and Y is already Galoisian over S. The rest of the assertion follows from 2.1.

Q.E.D.

In a similar vein to 2.7 we have

2.8. THEOREM. *Let A be a factorial k-algebra (UFD). Assume that A has no nontrivial units* $(A^* = k^*)$ *and k is of characteristic 3. Let B be a simple extension of A:*

$$B = A[X]/(P(X)) \quad \textit{for some} \quad P(X) \in A[X].$$

Then, if the natural map $\operatorname{spec} B \to \operatorname{spec} A$ *is étale finite of degree 3,* B *must be an Artin-Schreier extension of* A.

Proof. Writing $P(X) = X^3 + aX^2 + bX + c \in A[X]$, we will see what our assumptions bear on the coefficients a, b and c.

(i) By the unramifiedness of B over A, we have $(a,b) = 1$ clearly, where (x,y) denotes the g.c.d. of $x, y \in A$.

(ii) If $a = 0$, then $b \in k^*$ by (i) and B is of Artin-Schreier type over A. In fact, putting $\beta := \sqrt{b} \in k^*$, we see that $X^3 + bX + c = 0$ is equivalent to $(X/\beta)^3 + (X/\beta) + (c/\beta^3) = 0$.

(iii) Express the condition that $P(X)$ and $P'(X)$ generate the unit ideal in $A[X]$ (which is precisely the unramifiedness condition) as follows: $P'(X) = 2aX + b = -aX + b$, so $P'(X) = 0$ iff $X = b/a$. Since A is factorial and $P(b/a) = (b^3 - a^2b^2 + a^3c)/a^3$, the said condition is expressible as $b^3 - a^2b^2 + a^3c = \lambda \in A^* = k^*$ or, if we put $\alpha = \lambda^{1/3}$, as

$$(1) \qquad (b-\alpha)^3 = a^2(b^2 - ac) \qquad \text{for some} \quad \alpha \in k^* .$$

(iv) Now, if $a \in A^* = k^*$, then we may assume $a = 1$, for $X^3 + aX^2 + bX + c = a^3((X/a)^3 + (X/a)^2 + (b/a^2)(X/a) + (c/a^3))$. But, then, (1) becomes $(b-\alpha)^3 = b^2 - c$, so $c = -b^3 + b^2 + \alpha^3$; and then $X^3 + X^2 + bX + (-b^3 + b^2 + \alpha^3) = (X-b)^3 + (X-b)^2 + \alpha^3$, which is not irreducible. So, $a \in A^*$ is ruled out.

(v) By (ii) and (iv) we are left with only the case $a \in A$, $a \notin A^*$ and $a \neq 0$. For any prime element $s \in A$ that divides a, suppose $s^e \| a$ with $e > 0$ (meaning s^e divides a but s^{e+1} doesn't). Then, by (i) and (1), $s^{2e} \| (b-\alpha)^3$ and $s^q \| b - \alpha$ for some $q > 0$. So, $2e = 3q$, and $2 | q$, $3 | e$. One can thus rewrite

$$a = d^3, \quad b - \alpha = d^2 f \quad \text{with} \quad (d,f) = 1, \quad d \notin A^*.$$

Feeding these back into (1) one gets $d^6 f^3 = d^6((d^2f + \alpha)^2 - d^3c)$, or

$$(2) \qquad (f+\gamma)^3 = d^2(d(df^2 - c) - \alpha f)$$

with $\alpha \in k^*$ and $\gamma := (-\alpha^2)^{1/3} \in k^*$. But by the next lemma 2.9, an equation like (2) has no solutions $d \in A$, $f \in A$, $d \notin A^*$ with $(d,f) = 1$. This brings us back to (ii) above as the only feasible case. Q.E.D.

The following lemma, or at any rate its proof by the "method of infinite descent", might be of some interest on its own:

2.9. LEMMA. Let A be as in 2.8. For any $P \in A[x,y]$, $\mu \in k^*$ and $\nu \in k^*$ given, the equation

$$(3) \qquad (y+\mu)^3 = x^2(xP + \nu y)$$

has no solutions $x \in A$, $y \in A$ such that $x \notin A^*$ and $(x,y) = 1$.

Proof. Assume a pair $x \in A$ and $y \in A$ were a solution of (3) with $x \notin A^*$, $(x,y) = 1$. Then, as in step (v) of the proof of 2.8, one can find $z \in A$, $w \in A$ such that

$$x = z^3,\ y + \mu = z^2 w,\ z \notin A^* \text{ and } (z,w) = 1.$$

Substitute these back in (3) and we have $z^6 w^3 = z^6(z^3 P_1 + \nu(z^2 w - \mu))$ where $P_1 = P(z^3, z^2 w - \mu) \in A[z,w]$. Simplifying and shifting terms, we obtain

$$(4) \qquad (w+\lambda)^3 = z^2(zP_1 + \nu w)$$

where $\lambda = (\nu\mu)^{1/3} \in k^*$. This is of the same pattern as (3). So, by repeating the argument above on (4), we get $z = z_1^3$, $w + \lambda = z_1^2 w_1$ for $z_1 \in A$, $w_1 \in A$, so that $x = z^3 = z_1^9$ in particular. This process could go on ad infinitum allowing $x = z_1^9 = z_2^{27} = z_3^{81} = \ldots$, which is an absurdity. So no solutions as above could exist for (3).

Q.E.D.

2.10. Remark. Earlier, David Harbater suggested to us that the following assertion might be true: The Galois group of an étale Galois cover of $\mathbb{A}_k^1$ has always a normal subgroup of index $p = \mathrm{char}(k)$. He outlined to us a proof based on ideas contained in his thesis, and we mentioned this in our conference talk. Unfortunately, it has now become apparent that that outline is not readily workable into a complete proof, even though the assertion itself remains plausible. Anyhow, if the assertion is indeed true, then we can prove that an étale, degree p cover of $\mathbb{A}_k^1$ is necessarily Galoisian (hence of Artin-Shreier type). Let $Y \to \mathbb{A}^1$ be such a cover, and take the Galois closure $\tilde{Y} \to Y \to \mathbb{A}^1$ as before. The Galois group G of $\tilde{Y}$ over $\mathbb{A}^1$ is a subgroup of the symmetric group S_p on p letters. Let H be the subgroup corresponding to Y (i.e., $\tilde{Y}/H \simeq Y$), and N a normal subgroup of index p (as asserted). It is then an easy exercise in group theory to show $H = N$. Therefore, Y is Galoisian over $\mathbb{A}^1$.

REFERENCES

[1] Abhyankar, S. S. "Coverings of algebraic curves," Amer. J. Math., 79(1957), 825-856.

[2] Grothendieck, A. "Étude locale des schémas et des morphismes des schémas," Chap. IV, Part 4, Éléments de Géoémtrie Algébrique. No. 32 (1967), Publ. Math. I.H.E.S. (cited as EGA IV).

[3] Milne, J. S. Étale Cohomology Vol. 33 (1980), Princeton Mathematical Series. Princeton Univ. Press.

[4] Miyanishi, M. "p-Cyclic coverings of the afine space," J. Algebra 63 (1980) 279-284.

[5] Serre, J. P. "Espaces fibrés algébriques," exposé 1, Séminaire C. Chevalley, E.N.S.-Paris 1958.

[6] Villamayor, O. E., "Separable algebras and Galois extensions," Osaka Math. J. 4(1967), 161-171.

Current Address of the Authors:

T. Kambayashi
Department of Mathematical Sciences
Northern Illinois University
DeKalb, IL 60115, U.S.A.

V. Srinivas
School of Mathematics
Institute for Advanced Study
Princeton, NJ 08540, U.S.A.

ON SUPERSINGULAR ABELIAN VARIETIES

Niels O. Nygaard

Recall that an abelian variety A/k over a perfect field of characteristic $p > 0$ is said to be supersingular if the following three equivalent conditions are satisfied:

i) The p-divisible group $A[p]$ associated to A is isogenous to a product of copies of the p-divisible group $G_{1,1}$.

ii) All slopes of frobenius on the Dieudonne module $\mathbb{D}(A[p])$ are equal to 1/2.

iii) A is isogenous to a product of supersingular elliptic curves.

(The equivalence of i) and ii) is clear, the equivalence with iii) is due to Oort [8])

Recall further that the p-rank of A is defined as the rank of the unit root subcrystal of $\mathbb{D}(A[p])$ or equivalently as the length of the slope 0 part of the Newton polygon of $\mathbb{D}(A[p])$.

Let V denote the Cartier operator

$$V:\ H^0(A,\Omega^1_{A/k}) \longrightarrow H^0(A,\Omega^1_{A/k})$$

and let C be the matrix of V in some basis for the differentials of the first kind. If $C^{(\sigma)}$ denotes the matrix obtained by applying the frobenius to all the entries in C then it is well-known that we have

$$\text{p-rank}(A) = \text{rank}\ C^{(\sigma^{-(g-1)})}C^{(\sigma^{-(g-2)})}\cdots C^{(\sigma^{-1})}C$$

where $g = \dim(A)$. In particular p-rank$(A) = 0$ if and only if the matrix product above is zero. In the case $g = 1$ the constant C(1x1 matrix) is known as the Hasse invariant and an elliptic curve is supersingular if and only if C = the Hasse invariant is zero.

In the higher dimensional case it is no longer true that A is supersingular if its p-rank is zero so the matrix C does not contain enough information to decide supersingularity.

The purpose of this paper is to introduce a finite set of gxg matrices with entries in k, whose vanishing is equivalent to A being supersingular. These "higher Cartier-Manin" matrices can actually be computed in specific cases, particularly when A is the Jacobian of a curve. We consider the example of a hyper-elliptic curve and compute the higher Cartier-Manin matrices in terms of the coefficients of the defining equation.

1. The Main Theorem.

The main tool we shall use is Katz' "sharp-slope estimate" which we first recall:

1.1 Theorem (Katz [3]). Let (M,F) be an F-crystal and let $\lambda \geq 0$ be a rational number. Let $h^0, h^1, \ldots$ be the abstract Hodge numbers of (M,F). Then all the slopes in the Newton polygon are $\geq \lambda$ if and only if for all $n \geq 1$ we have:

$$F^{n+\sum_{i<\lambda} h^i} \text{ is divisible by } p^{\{n\lambda\}}$$

where $\{n\lambda\}$ denotes the closest integer $\geq n\lambda$.

Proof: See [3] 1.5.1.

1.2 Theorem. Let A/k be an abelian variety of dimension g, and consider the Dieudonne module $\mathbb{D}(A[p])$; all the slopes are in $[0,1]$, hence there is a σ^{-1}-linear map $V: \mathbb{D}(A[p]) \longrightarrow \mathbb{D}(A[p])$ such that $FV = VF = p$. A is supersingular if and only if

a) V^{g^2-g+2} is divisible by $p^{\frac{g^2+1}{2}-(g-1)}$ if g is odd

b) V^{g^2-2g+3} is divisible by $p^{\frac{g^2+1}{2}-\frac{3}{2}(g-1)}$ if g is even.

Proof: Let A^t be the dual abelian variety then there is a perfect pairing:

$$\mathbb{D}(A[p]) \times \mathbb{D}(A^t[p]) \longrightarrow W(k).$$

Under this pairing V_A is dual to F_{A^t}; since an abelian variety is supersingular if an only if its dual is, it will be enough to show that A is supersingular if and only if the corresponding divisibility properties hold for the same powers of F_A.

Now A is supersingular if and only if all the slopes on $\mathbb{D}(A[p])$ are $\geq 1/2$ and hence by 1.1 if and only if F^{n+h^o} is divisible by $p^{\{n/2\}}$ for all $n \geq 1$.

By Mazur and Messing [ ] we know that $\mathbb{D}(A[p]) = H^1_{crys}(A/W)$ as F-crystals and by Mazur's and Ogus' theorem ([3],[ ]) the abstract Hodge numbers of $H^1_{crys}(A/W)$ are equal to the Hodge numbers of A so $h^o = g$.

$\mathbb{D}(A[p])$ is a free W-module of rank $2g$, and since the break points of the Newton polygon occur at points with integer coordinates and since the Newton polygon is symmetric in the sense that if λ is a slope with multiplicity m then $1-\lambda$ is also a slope with multiplicity m (this last fact is a consequence of the strong Lefshetz theorem in crystalline cohomology) we see that the first possibility for a Newton polygon strictly below the line with slope 1/2 is as follows:

$$\text{g odd:} \quad \begin{cases} \text{slope } \frac{g-1}{2g} \text{ on the interval } [0,g] \\ \text{slope } \frac{g+1}{2g} \text{ on the interval } [g,2g] \end{cases}$$

$$\text{g even:} \quad \begin{cases} \text{slope } \frac{g-2}{2(g-1)} \text{ on the interval } [0,g-1] \\ \text{slope } 1/2 \quad \text{ on the interval } [g-1,g+1] \\ \text{slope } \frac{g}{2(g-1)} \text{ on the interval } [g+1,2g]. \end{cases}$$

Let $h^o(m), h^1(m), \ldots$ be the abstract Hodge numbers of F^m and let $\frac{1}{m}\text{Hodge}(F^m)$ be the Hodge polygon of these numbers with all slopes divided by m. By Katz [3] we know that

$$\frac{1}{m}\text{Hodge}(F^m) \leq \text{Newton}(F)$$

for all $m \geq 1$, and

$$\frac{1}{n}\text{Hodge}(F^m) \longrightarrow \text{Newton}(F) \quad m \to \infty.$$

It follows that the first (and hence all) Newton slope is 1/2 if and only if there is an integer m such that

$$\text{first slope in } \frac{1}{m}\text{Hodge}(F^m) > \begin{cases} \frac{g-1}{2g} & \text{g odd} \\ \frac{g-2}{2(g-1)} & \text{g even .} \end{cases}$$

Let n be an integer such that

$$\frac{\{n/2\}}{n+g} > \begin{cases} \frac{g-1}{2g} & \text{g odd} \\ \frac{g-2}{2(g-1)} & \text{g even.} \end{cases}$$

If F^{n+g} is divisible by $p^{\{n/2\}}$ then the first slope in $\frac{1}{n+g}\text{Hodge}(F^{n+g})$ satisfies:

$$\text{first slope} \geq \frac{\{n/2\}}{n+g} > \begin{cases} \frac{g-1}{2g} & \text{g odd} \\ \frac{g-2}{2(g-1)} & \text{g even} \end{cases}$$

and hence all the Newton slopes are 1/2. On the other hand if all the Newton slopes are 1/2 then by the sharp slope estimate F^{m+g} is divisible by $p^{\{m/2\}}$ for all $m \geq 1$.

If we take n to be the smallest integer satisfying the above inequalities we find

$$n = g^2-2g+2 \text{ and } \{n/2\} = \frac{g^2+1}{2} - (g-1) \qquad g \text{ odd}$$

$$n = g^2-2g+3 \text{ and } \{n/2\} = \frac{g^2+1}{2} - \frac{3}{2}(g-1) \qquad g \text{ even.}$$

It follows that A is supersingular if and only if

$$F^{g^2-g+2} \text{ is divisible by } p^{\frac{g^2+1}{2}-(g-1)} \qquad g \text{ odd}$$

$$F^{g^2-2g+3} \text{ is divisible by } p^{\frac{g^2+1}{2}-\frac{3}{2}(g-1)} \qquad g \text{ even.}$$

1.3 Corollary. A is supersingular if and only if

$$p^n \text{ divides } V^{2n-1+g} \text{ for } 1 \leq n \leq \begin{cases} \frac{g^2+1}{2}-(g-1) & g \text{ odd} \\ \frac{g^2+1}{2}-\frac{3}{2}(g-1) & g \text{ even.} \end{cases}$$

Proof: The necessity follows from the sharp slope estimate, sufficiency is clear by 1.2.

Since A is an abelian variety the spectral sequence

$$E_1^{i,j} = H^j(A,\Omega^i_{A/k}) \Longrightarrow H^{i+j}_{DR}(A/k)$$

degenerates at E_1 (e.g. [ ]); in particular we have an exact sequence

$$*)\qquad 0 \longrightarrow H^0(A,\Omega^1_{A/k}) \longrightarrow H^1_{DR}(A/k) \longrightarrow H^1(A,\mathcal{O}_A) \longrightarrow 0.$$

We have $H^i_{crys}(A/W) = \bigwedge^i H^1_{crys}(A/W)$ so all the crystalline cohomology is torsion free hence by the universal coefficient sequence we have $H^i_{crys}(A/W)\otimes k = H^i_{DR}(A/k)$ for all i; in particular for i = 1

$$\mathbb{D}(A[p])\otimes k = H^1_{crys}(A/W)\otimes k = H^1_{DR}(A/k).$$

Let L be a direct summand in $\mathbb{D}(A[p])$ lifting $H^0(A,\Omega^1_{A/k})$ and let M be a complement i.e. $L\oplus M = \mathbb{D}(A[p])$ and the exact sequence

$$0 \longrightarrow L \longrightarrow \mathbb{D}(A[p]) \longrightarrow M \longrightarrow 0$$

reduces to *) mod p.

1.4 Lemma. The matrix of V corresponding to the splitting

$$\mathbb{D}(A[p]) = L\oplus M$$

has the form

$$\begin{Bmatrix} X & Y \\ pZ & pU \end{Bmatrix}$$

where X, Y, Z, U are gxg matrices with entries in W.

Proof: Since FV = p we have Im $V \subset F^{-1}(p\mathbb{D}(A[p]))$ and if $x \in F^{-1}(\mathbb{D}(A[p]))$ we have $Fx = py$ so $x = pF^{-1}y = Vy$ (in $\mathbb{D}(A[p])\otimes\mathbb{Q}$) and hence Im V $= F^{-1}(p\mathbb{D}(A[p]))$. By Mazur's and Ogus' theorem the reduction of $F^{-1}(p\mathbb{D}(A[p]))$ is precisely $H^0(A,\Omega^1_{A/k})$ hence the reduction of V maps $H^1_{DR}(A/k)$ into $H^0(A,\Omega^1_{A/k})$.

Let

$$\begin{Bmatrix} X & Y \\ Z' & U' \end{Bmatrix}$$

be the matrix of V and let $x = \begin{Bmatrix} u \\ v \end{Bmatrix} \in L\oplus M = \mathbb{D}(A[p])$ then

$$Vx = \begin{Bmatrix} Xu + Yv \\ Z'u + U'v \end{Bmatrix}.$$

It follows that $Z'u + U'v$ reduces to 0 in $H^1_{DR}(A/k)$ for all u, v so Z' and U' are divisible by p.

1.5 Lemma. Assume $V^r|_L$ is divisible by p^m then $V^{r+i}L \subset p^mL + p^{m+1}M$ for all $i \geq 1$.

Proof: Let

$$\begin{Bmatrix} R_1 & R_2 \\ R_3 & R_4 \end{Bmatrix}$$

be the matrix of V^r then by the assumption $p^m|R_1$ and $p^m|R_3$.

The matrix of V^{r+1} is given by

$$\begin{Bmatrix} X & Y \\ pZ & pU \end{Bmatrix}\begin{Bmatrix} R_1^{\sigma^{-1}} & R_2^{\sigma^{-1}} \\ R_3^{\sigma^{-1}} & R_4^{\sigma^{-1}} \end{Bmatrix} = \begin{Bmatrix} XR_1^{\sigma^{-1}} + YR_3^{\sigma^{-1}} & XR_2^{\sigma^{-1}} + YR_4^{\sigma^{-1}} \\ pZR_1^{\sigma^{-1}} + pUR_3^{\sigma^{-1}} & pZR_2^{\sigma^{-1}} + pUR_4^{\sigma^{-1}} \end{Bmatrix}.$$

Since $p^m|XR_1^{\sigma^{-1}} + YR_3^{\sigma^{-1}}$ and $p^{m+1}|pZR_1^{\sigma^{-1}} + pUR_3^{\sigma^{-1}}$ the lemma follows by an obvious induction argument.

1.6 Definition. Let $\{\mu_1,\mu_2,\ldots,\mu_g\}$ be a basis of $H^0(A,\Omega^1_{A/k}) = L\otimes k$. Assume $V^r|_L$ is divisible by p^m then we define (m+1,r+i) Cartier-Manin matrix as follows:

By 1.5 we get a map

$$\frac{1}{p^m}V^{r+i}: L\otimes k \longrightarrow L\otimes k$$

the matrix C(m+1,r+i) of this map in the chosen basis is the (m+1,r+i) Cartier-Manin matrix.

<u>1.7 Lemma</u>. $V^{r+i}|_L$ is divisible by p^{m+1} if an only of $C(m+1,r+i) = 0$.

<u>Proof:</u> This is clear.

Remark that if $C(m+1,r+i) = 0$ then we can define $C(m+2,r+i+j)$ for all $j \geq 1$.

<u>1.8 Theorem.</u> A is supersingular if and only if

$$0 = C(1,g) = \ldots = C(r,g+2(r-1)) = \ldots = C\left(\frac{g^2+1}{2}-(g-1),\, g+2\left(\frac{g^2+1}{2}-g\right)\right),\quad g \text{ odd}$$

$$0 = C(1,g) = \ldots = C(r,g+2(r-1)) = \ldots = C\left(\frac{g^2+1}{2}-\frac{3}{2}(g-1),\, g+2\left(\frac{g^2+1}{2}-\frac{3}{2}(g-1)-1\right)\right),\quad g \text{ even.}$$

<u>Proof:</u> Remark first that the statement makes sense since if $C(r,g+2(r-1)) = 0$, $C(r+1,g+2r)$ is defined.

Assume the matrices in question vanish; by 1.3 we have to show that

$$p^n \mid V^{2n-1+g} \quad 1 \leq n \leq \begin{cases} \frac{g^2+1}{2}-(g-1) & g \text{ odd} \\ \frac{g^2+1}{2}-\frac{3}{2}(g-1) & g \text{ even} \end{cases}$$

$n = 1$: Since $C(1,g)$ vanishes $V^g|_L$ is divisible by p. Hence the matrix of V^g has the form

$$\begin{Bmatrix} pS_1 & S_2 \\ pT_1 & T_2 \end{Bmatrix}$$

so the matrix of V^{g+1} is

$$\begin{Bmatrix} pS_1 & S_2 \\ pT_1 & T_2 \end{Bmatrix} \begin{Bmatrix} X^{\sigma^{-1}} & Y^{\sigma^{-1}} \\ pZ^{\sigma^{-1}} & pU^{\sigma^{-1}} \end{Bmatrix} = \begin{Bmatrix} pS_1X^{\sigma^{-1}} + pS_2Z^{\sigma^{-1}} & pS_1Y^{\sigma^{-1}} + pS_2U^{\sigma^{-1}} \\ pT_1X^{\sigma^{-1}} + pT_2Z^{\sigma^{-1}} & pT_1Y^{\sigma^{-1}} + pT_2U^{\sigma^{-1}} \end{Bmatrix}$$

hence $p \mid V^{g+1}$.

Assume now that

$$p^n \mid V^{2n-1+g} \quad 1 \leq n \leq m < \begin{cases} \frac{g^2+1}{2}-(g-1) & g \text{ odd} \\ \frac{g^2+1}{2}-\frac{3}{2}(g-1) & g \text{ even} \end{cases}$$

we want to show

$$p^{m+1} \mid V^{2m+1+g}.$$

Since $C(m+1,g+2m) = 0$ and since $p^m \mid V^{2m-1+g}$ and hence also $p^m \mid V^{2m+g}$ the matrix of V^{2m+1+g} is:

$$\begin{Bmatrix} p^{m+1}R_1 & p^m R_2 \\ p^{m+1}R_3 & p^m R_4 \end{Bmatrix} \begin{Bmatrix} X^{\sigma^{-1}} & Y^{\sigma^{-1}} \\ pZ^{\sigma^{-1}} & pU^{\sigma^{-1}} \end{Bmatrix}$$

$$= \begin{Bmatrix} p^{m+1}R_1X^{\sigma^{-1}} + p^{m+1}R_2Z^{\sigma^{-1}} & p^{m+1}R_1Y^{\sigma^{-1}} + p^{m+1}R_2U^{\sigma^{-1}} \\ p^{m+1}R_3X^{\sigma^{-1}} + p^{m+1}R_4Z^{\sigma^{-1}} & p^{m+1}R_3Y^{\sigma^{-1}} + p^{m+1}R_4U^{\sigma^{-1}} \end{Bmatrix}.$$

This proves that $p^{m+1} \mid V^{2m+1+g}$ and by induction it follows that A is supersingular.

Assume next that A is supersingular then $p^n \mid V^{2n-1+g}$ for all $n \geq 1$. Since L reduces mod p to $H^0(A,\Omega^1_{A/k})$ it follows from Mazur's and Ogus' theorem that $L \subset F^{-1}(p\mathbb{D}(A[p])) = \operatorname{Im} V$ hence

$$V^{2(n-1)+g}(L) \subset V^{2(n-1)+g}(\operatorname{Im} V) \subset \operatorname{Im} V^{2n-1+g} \subset p^n\mathbb{D}(A[p]) \text{ all } n \geq 1$$

and so all the matrices

$$C(1,g),\ C(2,g+2), \ldots,\ C(n,g+2(n-1)), \ldots$$

vanish.

In this set-up we have looked at V rather than F (this is mainly for convenience in the example we consider in the next section), but it is clear that we have the statements analogous to 1.5, 1.6 and 1.7 for powers of F restricted to M. The matrices thus obtained could be called the higher Hasse-Witt matrices; we denote them by $H(n,i)$. It is clear that we have the following criterion for supersingularity in terms of the higher Hasse-Witt matrices.

<u>1.9 Theorem.</u> A is supersingular if and only if the matrices $H(n,g+2(n-1))$ vanish for

$$1 \leq n \leq \begin{cases} \frac{g^2+1}{2} - (g-1) & g \text{ odd} \\ \frac{g^2+1}{2} - \frac{3}{2}(g-1) & g \text{ even.} \end{cases}$$

2. Hyperelliptic curves.

Let $X/W(k)$ be a hyperelliptic curve with affine equation

$$u^2 = t^{2g+1} + a_{2g}t^{2g} + a_{2g-1}t^{2g-1} + \ldots + a_1 t + a_0 = F(t)$$

we assume $p \neq 2$, $a_0 \not\equiv 0 \bmod p$ and $a_0^{\frac{1}{2}} \in W(k)$.

A basis of differentials of the first kind is given by

$$\frac{dt}{u},\ t\frac{dt}{u}, \ldots\ldots,\ t^{g-1}\,\frac{dt}{u} \in H^0(X, \Omega^1_{X/W(k)}).$$

Let $c(r,n)$ denote the coefficient to t^r in the polynomial $F(t)^{\frac{p^n-1}{2}}$ then we have the following criterion for supersingularity of the Jacobian $J(X_o)$ where X_o is the curve $X \otimes k$.

2.1 Theorem. $J(X_o)$ is supersingular if and only if the matrix

$$\left\{c(ip^{g+2(n-1)}-j, g+2(n-1))\right\}_{\substack{i=1,\ldots,g\\ j=1,\ldots,g}} \equiv 0 \bmod p^n$$

for

$$1 \leq n \leq \begin{cases} \frac{g^2+1}{2} - (g-1) & g \text{ odd} \\[1ex] \frac{g^2+1}{2} - \frac{3}{2}(g-1) & g \text{ even} \end{cases}$$

We shall prove this theorem through a series of lemmas. I should like to thank B. Dwork for some very helpful conversations concerning the computations of this section.

2.2 Lemma. Let $f(t) \in W[[t]]$ then there are powerseries $f_1(t), f_2(t), \ldots$ in $W[[t]]$ such that for all $m \geq 0$ we have

$$f(t)^{p^m} = f^{\sigma^m}(t^{p^m}) + pf_1^{\sigma^{m-1}}(t^{p^{m-1}}) + \ldots\ldots + p^m f_m(t)$$

($g^{\sigma}(t)$ denotes the powerseries obtained by applying σ to all the coefficients of $g(t)$). If $f(t)$ is a polynomial the $f_1(t), f_2(t), \ldots$ can be chosen to be polynomials as well.

Proof: Define $f_r(t) = \frac{1}{p^r}(f(t)^{p^r} - f\ (t^p)^{p^{r-1}})$ then we have:

$$f^{\sigma^m}(t^{p^m}) + pf_1^{\sigma^{m-1}}(t^{p^{m-1}}) + p^2 f_2^{\sigma^{m-2}}(t^{p^{m-2}}) + \ldots + p^m f_m(t)$$

is equal to:

$$f^{\sigma^m}(t^{p^m}) + (f^{\sigma^{m-1}}(t^{p^{m-1}})^p - f^{\sigma^m}(t^{p^m})) + (f^{\sigma^{m-2}}(t^{p^{m-2}})^{p^2} - f^{\sigma^{m-1}}(t^{p^{m-1}})^p)$$

$$+\ldots\ldots+ (f(t)^{p^m} - f^{\sigma}(t^p)^{p^{m-1}}) = f(t)^{p^m}$$

so it remains to show that $f_m(t) \in W[[t]]$.

Assume that

$$f(t)^{p^{r-1}} \equiv f^{\sigma}(t^p)^{p^{r-2}} \mod p^{r-1}$$

for some $r \geq 2$, then

$$f(t)^{p^{r-1}} = f^{\sigma}(t^p)^{p^{r-2}} + p^{r-1}g(t)$$

for some $g(t) \in W[[t]]$. It follows that

$$\begin{aligned} f(t)^{p^r} &= (f^{\sigma}(t^p)^{p^{r-2}} + p^{r-1}g(t))^p \\ &= f^{\sigma}(t^p)^{p^{r-1}} + \tbinom{p}{1}p^{r-1}g(t)f^{\sigma}(t^p)^{p^{r-2}(p-1)} \\ &+ \ldots\ldots + \tbinom{p}{p-1}(p^{r-1}g(t))^{p-1}f^{\sigma}(t^p)^{p^{r-2}} \\ &+ (p^{r-1}g(t))^p \\ &\equiv f^{\sigma}(t^p)^{p^{r-1}} \mod p^r. \end{aligned}$$

Hence it is enough to show

$$f(t)^p \equiv f^{\sigma}(t^p) \mod p$$

which is immediate.

It is clear that if $f(t)$ is a polynomial then the $f_r(t)$ are also polynomials.

2.3 Lemma. Let $c(r,n)$ be defined as above. Assume that

$$\left\{c(ip^{g+2(m-1)}-j, g+2(m-1))\right\}_{\substack{i=1,\ldots,g \\ j=1,\ldots,g}} \equiv 0 \mod p^m$$

for $m \leq n$ then we have

a) $p^n \,|\, c(ip^{g+2(n-1)}-j, g+2(n-1))$ $\qquad 1 \leq j \leq g$

$p^{n-1} \,|\, c(ip^{g+2(n-1)-1}-j, g+2(n-1))$ $\qquad \vdots$

$\vdots$

$p \,|\, c(ip^{g+n-1}-j, g+2(n-1))$

$$\text{b)}\qquad p^n \mid c(ip^{g+2(n-1)}-j,\, g+2n-1) \qquad 1 \le i,j \le g$$

$$p^{n-1} \mid c(ip^{g+2(n-1)-1}-j,\, g+2n-1) \qquad .$$

$$\vdots \qquad\qquad \vdots$$

$$p \mid c(ip^{g+n-1}-j,\, g+2n-1) \qquad .$$

$$\text{c)}\qquad p^n \mid c(ip^{g+2n-1}-j,\, g+2n-1)$$

Proof: a): Suppose first that $n = 1$ then the statement is a tautology. Assume next that a) holds up to $n-1$, so we have

$$p^{n-1} \mid c(ip^{g+2(n-2)}-j,\, g+2(n-2)) \qquad 1 \le i,j \le g$$

$$p^{n-2} \mid c(ip^{g+2(n-2)-1}-j,\, g+2(n-2)) \qquad .$$

$$\vdots \qquad\qquad \vdots$$

$$p \mid c(ip^{g+n-2}-j,\, g+2(n-2)) \qquad .$$

Write $F(t)^{\frac{p^{g+2(n-1)}-1}{2}} = F(t)^{\frac{p^{g+2(n-2)}-1}{2}} \left(F(t)^{\frac{p^2-1}{2}}\right)^{p^{g+2(n-2)}}$

by 2.2 there are polynomials $G_0(t), G_1(t), \ldots \left(G_0(t) = \left(F^{\sigma^{g+2(n-2)}}(t)\right)^{\frac{p^2-1}{2}}\right)$ such that

$$\left(F(t)^{\frac{p^2-1}{2}}\right)^{p^{g+2(n-2)}} = G_0\left(t^{p^{g+2(n-2)}}\right) + pG_1\left(t^{p+2(n-2)-1}\right) + \ldots\ldots + p^{g+2(n-2)}G_{g+2(n-2)}(t).$$

Let $1 \le k \le n-1$ then we have the following congruence:

$$F(t)^{\frac{p^{g+2(n-1)}-1}{2}} \equiv F(t)^{\frac{p^{g+2(n-2)}-1}{2}} \left(G_0\left(t^{p^{g+2(n-2)}}\right) + pG_1\left(t^{p^{g+2(n-2)-1}}\right) + \ldots + p^{n-k-1}G_{n-k-1}\left(t^{p^{g+n-2+k+1}}\right)\right) \bmod p^{n-k}.$$

If we let $G_d(t) = b_s(d)t^d$ then we have the following congruences between the coefficients:

$$c(r, g+2(n-1)) \equiv \sum_{u+sp^{g+2(n-2)}=r} c(u, g+2(n-2))b_s(0) + p \sum_{u+sp^{g+2(n-2)-1}=r} c(u, g+2(n-2))b_s(1) + \ldots + p^{n-k-1} \sum_{u+sp^{g+n+k-1}=r} c(u, g+2(n-2))b_s(n-k-1) \quad \bmod p^{n-k}.$$

Now for $r = ip^{g+2(n-1)-k}-j$ we have

$$u + sp^{g+2(n-2)} = ip^{g+2(n-1)-k}-j \Rightarrow u = (pi-p^{k-1}s)p^{g+2(n-2)-(k-1)}-j$$

$$u + sp^{g+2(n-2)-1} = ip^{g+2(n-1)-k}-j \Rightarrow u = (p^2 i-p^{k-1}s)p^{g+2(n-2)-k}-j$$

$$\vdots \qquad \vdots \qquad \vdots \qquad \vdots$$

$$u + sp^{g+n+k-1} = ip^{g+2(n-1)-k}-j \Rightarrow u = (p^{n-k}i-p^{k-1})p^{g+n-2}-j$$

and by the induction hypothesis

$$p^{n-1-(k-1)} = p^{n-k} \mid c((pi-p^{k-1}s)p^{g+2(n-2)-k+1}-j,g+2(n-2))$$

$$p^{n-1-(k-1)-1} = p^{n-k-1} \mid c((p^2 i-p^{k-1}s)p^{g+2(n-2)-k}-j,g+2(n-2))$$

$$\vdots$$

$$p \mid c((p^{n-k}i-p^{k-1}s)p^{g+n-2}-j,g+2(n-2))$$

and so $p^{n-k} \mid c(ip^{g+2(n-1)-k}-j,g+2(n-1)) \quad 1 \le k \le n-1$.

The only left to prove in a) is that

$$p^n \mid c(ip^{g+2(n-1)}-j,g+2(n-1))$$

but that is part of the assumptions.

To prove b) and c), write

$$F(t)^{\frac{p^{g+2n-1}-1}{2}} = F(t)^{\frac{p^{g+2(n-1)}-1}{2}}(F(t)^{\frac{p^2-1}{2}})^{p^{g+2(n-1)}}$$

and proceed as above, b) and c) then follow from a).

2.4 Corollary. Under the assumptions of 2.3 we have

$$c(ip^{g+2n}-j,g+2n) \equiv \sum_k c(kp^{g+2n-1}-j,g+2n-1)c(ip-k,1)^{g+2n-1} \mod p^{n+1}$$

Proof: Write

$$F(t)^{\frac{p^{g+2n}-1}{2}} = F(t)^{\frac{p^{g+2n-1}-1}{2}}(F(t)^{\frac{p^2-1}{2}})^{p^{g+2n-1}}$$

$$\equiv F(t)^{\frac{p^{g+2n-1}-1}{2}}(F^{g+2n-1}(t^{p^{g+2n-1}})^{\frac{p-1}{2}} + pG_1(t^{p^{g+2n-2}})$$

$$+ \ldots + p^n G_{n-1}(t^{p^{g+n-1}})) \mod p^{n+1}.$$

In terms of coefficients we have:

$$c(r,g+2n) \equiv \sum_{u+sp^{g+2n-1}=r} c(u,g+2n-1)c(s,1)^{\sigma^{g+2n-1}}$$
$$+ p \sum_{u+sp^{g+2n-2}=r} c(u,g+2n-1)b_s(1)$$
$$\vdots$$
$$+ p^n \sum_{u+sp^{g+n-1}=r} c(u,g+2n-1)b_s(n) \quad \bmod p^{n+1}$$

If $r = ip^{g+2n}-j$ we have

$$u+sp^{g+2n-1} = ip^{g+2n}-j \quad \Rightarrow \begin{cases} u = p^{g+2n-1}(ip-s)-j \\ \text{and if } k=ip-s,\ s=ip-k \end{cases}$$

$$u+sp^{g+2n-2} = ip^{g+2n}-j \quad \Rightarrow \quad u = p^{g+2n-2}(ip^2-s)-j$$
$$\vdots \qquad\qquad \vdots \qquad\qquad \vdots$$
$$u+sp^{g+n-1} = ip^{g+2n}-j \quad \Rightarrow \quad u = p^{g+n-1}(ip^n-s)-j.$$

By 2.3

$$p^n \mid c((p^{g+2n-2}(ip^2-s)-j,g+2n-1)$$
$$\vdots$$
$$p \mid c(p^{g+n-1}(ip^n-s)-j,g+2n-1)$$

hence

$$c(ip^{g+2n}-j,g+2n) \equiv \sum_k c(p^{g+2n-1}k-j,g+2n-1)c(ip-k,1)^{\sigma^{g+2n-1}} \bmod p^{n+1}$$

2.5 Lemma. Let Y/W be a smooth and proper curve with a W-rational point y. Let $\hat{Y}_y$ be the formal completion at y and let $Y_o = Y \otimes k$.

Consider the formal expansion map

$$\mu:\ H^1_{crys}(Y_o/W) = H^1_{DR}(Y/W) \longrightarrow H^1_{DR}(\hat{Y}_y/W)$$

then $\mu^{-1}(pH^1_{DR}(\hat{Y}_y/W)) = \text{Im } F$.

Proof: Let μ_o denote the formal expansion map $H^1_{DR}(Y_o/k) \longrightarrow H^1_{DR}(\hat{Y}_y \otimes k/k)$ Katz [4] has shown that $\ker \mu_o = F_1^{con} H^1_{DR}(Y_o/k) = H^1(Y_o, \mathcal{H}^0(\Omega^{\bullet}_{Y_o/k}))$. In [4] it is also shown that $H^1_{DR}(Y_y/W)\otimes k \longrightarrow H^1_{DR}(Y_y \otimes k/k)$ is an injection (compare this with the universal coefficient sequence in crystalline cohomology). It is clear that μ_o factors through $H^1_{DR}(\hat{Y}_y/W)\otimes k$ so we have an exact sequence:

$$0 \longrightarrow H^1(Y_o, \mathcal{H}^0(\Omega^{\bullet}_{Y_o/k})) \to H^1_{DR}(Y_o/k) \longrightarrow H^1_{DR}(Y_y/W)\otimes k$$

Now $F_1^{con}H^1_{DR}(Y_o/k) = \mathrm{Im}\, F: H^1_{DR}(Y_o/k) \longrightarrow H^1_{DR}(Y_o/k)$, it follows that if $x \in \mu^{-1}(pH^1_{DR}(\hat{Y}_y/W))$ then $\bar{x} = F\bar{z}$ ($\bar{\ }$ denotes reduction mod p) hence $x \in \mathrm{Im}\, F + pH^1_{crys}(Y_o/W) \subset \mathrm{Im}\, F$.

The other inclusion is trivial since F is divisible by p on $H^1_{DR}(\hat{Y}_y/W)$ [4].

2.6 Lemma. Under the assumptions of 2.3

$$V^{g+2n}H^0(X,\Omega^1_{X/W}) \subset p^nH^0(X,\Omega^1_{X/W}) + p^{n+1}H^1_{crys}(X_o/W)$$

and the matrix of

$$V^{g+2n}: H^0(X,\Omega^1_{X/W})\otimes W_{n+1} \longrightarrow H^0(X,\Omega^1_{X/W})\otimes W_{n+1}$$

in the basis $\left\{\frac{dt}{u}, t\frac{dt}{u}, t^2\frac{dt}{u}, \ldots, t^{g-1}\frac{dt}{u}\right\}$ is given by

$$\left\{c(ip^{g+2n}-j,g+2n)^{\sigma^{-(g+2n)}}\right\}_{\substack{i=1,\ldots,g\\ j=1,\ldots,g}} .$$

Proof: We first compute the matrix of

$$V^g: H^0(X_o,\Omega^1_{X_o/k}) \longrightarrow H^0(X_o,\Omega^1_{X_o/k}).$$

Consider $\omega_j = t^{j-1}u^{-1}dt$ as a 1-form on the functionfield $k(X_o)$. We have

$$\omega_j = t^ju^{-1}\frac{dt}{t} = t^ju^{p^g-1}u^{-p^g}\frac{dt}{t} = t^jF(t)^{\frac{p^g-1}{2}}u^{-p^g}\frac{dt}{t}$$

$$= \sum_r c(r,g)t^{r+j}u^{-p^g}\ \frac{dt}{t}$$

so

$$V^g\omega_j = V^{g-1}\Big(\sum_r c(r,g)^{\sigma^{-1}}V(t^{r+j}u^{-p^g}\ \frac{dt}{t})\Big)$$

$$= V^{g-1}\Big(\sum_r c(r,p)^{\sigma^{-1}}u^{-p^{g-1}}V(t^{r+j}\ \frac{dt}{t})\Big).$$

If $p\nmid r+j$, $t^{r+j}\ \frac{dt}{t} = \frac{1}{r+j}\ d(t^{r+j})$ so $V(t^{r+j}\ \frac{dt}{t}) = 0$. If $r+j = pi_1$ we have $V(t^{pi_1}\frac{dt}{t}) = t^{i_1}\ \frac{dt}{t}$ so

$$V^g\omega_j = V^{g-1}\Big(\sum c(pi_1-j,g)^{\sigma^{-1}}u^{-p^{g-1}}t^{i_1}\frac{dt}{t}\Big)$$

$$= V^{g-2}\Big(\sum c(pi_1-j,g)^{\sigma^{-2}}u^{-p^{g-2}}V(t^{i_1}\frac{dt}{t})\Big)$$

$$= V^{g-2}\Big(\sum c(p^2i_2-j,g)^{\sigma^{-2}}u^{-p^{g-2}}t^{i_2}\frac{dt}{t}\Big)$$

$$\vdots$$

$$= \sum_{i_g} c(p^gi_g-j,g)^{\sigma^{-g}}u^{-1}t^{i_g}\frac{dt}{t} = \sum_i c(p^gi-j,g)^{\sigma^{-g}}\ \omega_i$$

This shows that the matrix of V^g is

$$\left\{c(p^g i-j,g)^{\sigma^{-g}}\right\}_{\substack{i=1,\dots,g\\ j=1,\dots,g}}$$

(The idea of computing the matrix this way is due to Manin [7])

Assume now that the lemma holds for $m \leq n-1$ then the matrix of $V^{g+2(n-1)}: H^0(X,\Omega^1_{X/W})\otimes W_n \longrightarrow H^0(X,\Omega^1_{X/W})\otimes W_n$ is given by

$$\left\{c(ip^{g+2(n-1)}-j,g+2(n-1))^{\sigma^{-(g+2(n-1))}}\right\}_{\substack{i=1,\dots,g\\ j=1,\dots,g}}$$

which by our assumption is divisible by p^n. It follows that the matrix of $V^{g+2(n-1)}$ on $H^1_{crys}(X_0/W)$ corresponding to the splitting $H^0(X,\Omega^1_{X/W})\oplus H^1(X,{}_X)$ has the form (see the proof of 1.8)

$$\begin{Bmatrix} p^nR_1 & p^{n-1}R_2 \\ p^nR_3 & p^{n-1}R_4 \end{Bmatrix}.$$

By 1.4 the matrix of V has the form

$$\begin{Bmatrix} A & B \\ pC & pD \end{Bmatrix}$$

so the matrix of V^{g+2n-1} is

$$\begin{Bmatrix} p^nR_1 & p^{n-1}R_2 \\ p^nR_3 & p^{n-1}R_4 \end{Bmatrix}\begin{Bmatrix} A & B \\ pC & pD \end{Bmatrix} = \begin{Bmatrix} p^nR_1A+p^nR_2C & p^nR_1B+p^nR_2D \\ p^nR_3A+p^nR_4C & p^nR_3B+p^nR_4D \end{Bmatrix}$$

which shows that V^{g+2n-1} is divisible by p^n. Now the matrix of V^{g+2n} is:

$$\begin{Bmatrix} A & B \\ pC & pD \end{Bmatrix}\begin{Bmatrix} p^n(R_1A+R_2C) & p^n(R_1B+R_2D) \\ p^n(R_3A+R_4C) & p^n(R_3B+R_4D) \end{Bmatrix}$$

$$= \begin{Bmatrix} p^n(AR_1A+AR_2C)+p^n(BR_3A+BR_4C) & p^n(AR_1B+AR_2D)+p^n(BR_3B+BR_4D) \\ p^{n+1}(CR_1A+CR_2C)+p^{n+1}(DR_3A+DR_4C) & p^{n+1}(CR_1B+CR_2D)+p^{n+1}(DR_3B+DR_4D) \end{Bmatrix}$$

(we have suppressed the σ^{-1}-linearity of V)
This expression proves the first part of the theorem.
Let x be the W(k)-rational point $(0,\sqrt{a_0})$, t is a local parameter at x, hence the elements of $H^1_{DR}(\hat{X}_x/W)$ can be represented in the form $g(t)\frac{dt}{t}$ with $g(t)\in tW(k)[[t]]$.
F and V act as follows:

$$F(g(t)\frac{dt}{t})=pg^{\sigma}(t^p)\frac{dt}{t}$$

$$V(g(t)\frac{dt}{t})=g^{\sigma^{-1}}(t^{1/p})\frac{dt}{t} \quad \text{where } t^{n/p}=0 \text{ if } p\nmid n$$

Let $u^{-1}=F(t)^{-1/2}=f(t)\in W(k)[[t]]$, then the formal expansion of ω_j at x is

$$\phi(\omega_j)=t^j f(t)\frac{dt}{t} = t^j F(t)^{\frac{p^{g+2n-1}-1}{2}} f(t)^{p^{g+2n-1}}\frac{dt}{t}$$

By (2.2) we have

$$f(t)^{p^{g+2n-1}}=f^{\sigma^{g+2n-1}}(t^{p^{g+2n-1}})+pf_1^{\sigma^{g+2n-2}}(t^{p^{g+2n-2}})+\ldots+p^{g+2n-1}f_{g+2n-1}(t)$$

hence we get $\phi(\omega_j)$

$$=t^j F(t)^{\frac{p^{g+2n-1}-1}{2}}\left(f^{\sigma^{g+2n-1}}(t^{p^{g+2n-1}})+pf_1^{\sigma^{g+2n-2}}(t^{p^{g+2n-2}})+\ldots\right.$$

$$\left.\ldots+p^{g+2n-1}f_{g+2n-1}(t)\right),$$

and writing out $F(t)^{\frac{p^{g+2n-1}-1}{2}}$ we have

$$\phi(\omega_j)=\sum c_r(g+2n-1)t^{j+r}f^{\sigma^{g+2n-1}}(t^{p^{g+2n-1}})\frac{dt}{t}$$

$$+p\sum c_r(g+2n-1)t^{j+r}f_1^{g+2n-2}(t^{p^{g+2n-2}})\frac{dt}{t}+\ldots$$

$$\ldots+p^{g+2n-1}\sum c_r(g+2n-1)t^{j+r}f_{g+2n-1}(t)\frac{dt}{t}$$

It follows that $\phi(V^{g+2n-1}\omega_j) = V^{g+2n-1}(\phi(\omega_j))$

$$=\sum c_r^{\sigma^{g+2n-1}}(g+2n-1)V^{g+2n-1}(t^{j+r}f^{\sigma^{g+2n-1}}(t^{g+2n-1})\frac{dt}{t}$$

$$+\ p\sum c_r^{\sigma^{g+2n-1}}(g+2n-1)V^{g+2n-1}(t^{j+r}f_1^{\sigma^{g+2n-2}}(t^{g+2n-2})\frac{dt}{t})$$

$$\vdots$$

$$+\ p^n\sum c_r^{\sigma^{g+2n-1}}(g+2n-1)V^{g+2n-1}(t^{j+r}f_n(t^{g+n-1})\frac{dt}{t})$$

$$+\ p^{n+1}\delta,\ \delta\in H^1_{DR}(X_x/W)$$

Now

$$V^{g+2n-1}(t^{j+r}f^{\sigma^{g+2n-1}}(t^{p^{g+2n-1}})\frac{dt}{t}) = \begin{cases} 0 & \text{if } p^{g+2n-1} \nmid j+r \\ t^i f(t)\frac{dt}{t} & \text{if } j+r=ip^{g+2n-1}\end{cases}$$

$$V^{g+2n-1}(t^{j+r}f_1^{g+2n-2}(t^{p^{g+2n-2}})\frac{dt}{t} = \begin{cases} 0 & \text{if } p^{g+2n-2} \nmid j+r \\ V(t^i f_1(t)\frac{dt}{t}) & \text{if } j+r=ip^{g+2n-2}\end{cases}$$

$$V^{g+2n-1}(t^{j+r}f_n^{g+n-1}(t^{p^{g+n-1}})\frac{dt}{t}) = \begin{cases} 0 & \text{if } p^{g+n-1} \nmid j+r \\ V^n(t^i f_n(t)\frac{dt}{t}) & \text{if } j+r=ip^{g+n-1}\end{cases}$$

so $\phi(V^{g+2n-1}\omega_j)$

$$= \sum c^{\sigma^{g+2n-1}}_{ip^{g+2n-1}}(g+2n-1)t^i f(t)\frac{dt}{t}$$

$$+ p\sum c^{\sigma^{g+2n-1}}_{ip^{g+2n-2}-j}(g+2n-1)V(t^i f_1(t)\frac{dt}{t})$$

$$\vdots$$

$$+ p^n\sum c^{\sigma^{g+2n-1}}_{ip^{g+n-1}-j}(g+2n-1)V^n(t^i f_n(t)\frac{dt}{t})$$

$$+ p^{n+1}\delta$$

By (2.3)

$$p^n \mid c_{ip^{g+2n-1}-j}(g+2n-1)$$

and $p^n \mid c_{ip^{g+2(n-1)}-j}(g+2n-1)$

$$\vdots$$

$$p \mid c_{ip^{g+n-1}-j}(g+2n-1)$$

so we get

$$\phi(\frac{1}{p^n}V^{g+2n-1}\omega_j - \sum\frac{1}{p^n}c^{\sigma^{g+2n-1}}_{ip^{g+2n-1}}(g+2n-1)\omega_i)$$

$$= \frac{1}{p^n}\phi(V^{g+2n-1}\omega_j) - \sum\frac{1}{p^n}c^{\sigma^{g+2n-1}}_{ip^{g+2n-1}-j}(g+2n-1)t^i f(t)\frac{dt}{t}$$

$$= p\sum\frac{1}{p^n}c^{\sigma^{g+2n-1}}_{ip^{g+2(n-1)}-j}(g+2n-1)V(t^i f_1(t)\frac{dt}{t})$$

$$+ \vdots$$

$$p\sum\frac{1}{p^n}p^{n-1}c^{\sigma^{g+2n-1}}_{ip^{g+n-1}-j}(g+2n-1)V^n(t^i f_n(t)\frac{dt}{t}) + p\delta$$

$$\varepsilon\, pH^1_{DR}(\hat{X}_x/W)$$

It follows now from (2.5) that

$$\frac{1}{p^n}V^{g+2n-1}\omega_j - \sum \frac{1}{p^n} c^{\sigma^{-(g+2n-1)}}_{ip^{g+2n-1}-j}(g+2n-1)\omega_j = F\alpha$$

for some $\alpha \in H^1_{crys}(x_o/W)$ and so

$$V^{g+2n}\omega_j - \sum c^{\sigma^{-(g+2n)}}_{ip^{g+2n-1}-j}(g+2n-1)\; V\omega_i = p^n VF\alpha = p^{n+1}\alpha$$

Now $V\omega_i \equiv \sum c^{\sigma^{-1}}_{pk-i}(1)\;_k \bmod p$

hence $\sum c^{\sigma^{-(g+2n)}}_{ip^{g+2n-1}-j}(g+2n-1)V\omega_i$

$$\equiv \sum_i \sum_k c^{\sigma^{-(g+2n)}}_{ip^{g+2n-1}-j}(g+2n-1)\; c^{\sigma^{-1}}_{pk-i}(1)\omega_k \bmod p^{n+1}$$

and by (2.4)

$$\sum_i c^{\sigma^{-(g+2n)}}_{ip^{g+2n-1}-j}(g+2n-1)c^{\sigma^{-1}}_{pk-1}(1) \equiv c^{\sigma^{-(g+2n)}}_{kp^{g+2n}-j}(g+2n) \bmod p^{n+1}$$

so

$$V^{g+2n}\omega_j \equiv \sum_k c^{\sigma^{-(g+2n)}}_{kp^{g+2n}-j}(g+2n)\omega_k \bmod p^{n+1}$$

which proves the second statement.

We can now easily finish the proof of theorem 2.1.

Remark first that $H^1_{crys}(X_o/W) = H^1_{crys}(J(X_o)/W)$ and

$H^o(X_o,\Omega^1_{X_o/k}) = H^o(J(X_o),\Omega^1_{J(X_o)/k})$, so $H^o(X,\Omega^1_{X/W})$

is a lifting of $H^o(J(X_o),\Omega^1_{J(X_o)/k})$ and we can compute the higher Cartier-Manin matrices from the action of V on $H^o(X,\Omega^1_{X/W})$.

Assume now that

$$\{c_{ip^{g+2(n-1)}-j}(g+2(n-1))\}_{\substack{i=1,\dots g\\ j=1,\dots g}} \equiv 0 \bmod p^n$$

$$\text{for } 1\le n\le \begin{cases} \frac{g^2+1}{2} - (g-1) & g \text{ odd}\\ \frac{g^2+1}{2} - \frac{3}{2}(g-1) & g \text{ even}\end{cases}$$

It follows immediately from (2.6) that

$$\mathbf{C}(n,g+2(n-1)) \equiv \frac{1}{p^{n-1}}\left\{c^{\sigma^{-(g+2(n-1))}}_{ip^{g+2(n-1)}-j}(g+2(n-1))\right\}_{\substack{i=1,\dots,g\\ j=1,\dots,g}} \mod p$$

hence $J(X_o)$ is supersingular by (1.8).

Assume next that $J(X_o)$ is supersingular. Then $V^g|_{H^o(X,\Omega^1_{X/W})}$ is divisible by p, hence by the computation of the matrix of V^g mod p we have

$$\left\{c_{ip^g-j}(g)\right\}_{\substack{i=1,\dots,g\\ j=1,\dots,g}} \equiv 0 \mod p$$

by (2.6) this implies that the matrix of

$$V^{g+2}: H^o(X,\Omega^1_{X/W}) \otimes W_2 \to H^o(X,\Omega^1_{X/W}) \otimes W_2$$

is given by

$$\left\{c^{\sigma^{-(g+2)}}_{ip^{g+2}-j}(g+2)\right\}_{\substack{i=1,\dots,g\\ j=1,\dots,g}}$$

and by the supersingularity $V^{g+2}|_{H^o(X,\Omega^1_{X/W})}$

is divisible by p^2, hence

$$\left\{c_{ip^{g+2}-j}(g+2)\right\}_{\substack{i=1,\dots,g\\ j=1,\dots,g}} \equiv 0 \mod p^2$$

Proceeding by induction we get

$$\left\{c_{ip^{g+2(n-1)}-j}(g+2(n-1))\right\}_{\substack{i=1,\dots,g\\ j=1,\dots,g}} \equiv 0 \mod p^n$$

for all $n \geq 1$.

REFERENCES

1. Demazure, M.-Lectures on p-divisible groups. Lecture Notes in Mathematics, No. 302, Springer Verlag, Berlin, Heidelberg, New York (1972).

2. Honda, T.-On the Jacobian variety of the algebraic curve $y^2=1-x^\ell$ over a field characteristic $p>0$. Osaka J. Math. 3, pp. 189-194 (1966)

3. Katz, N.-Slope filtration of F-crystals. Asterisque 64, 1979

4. Katz, N.-Crystalline Cohomology, Dieudonne Modules and Jacobi sums. In Automorphic forms, Representation theory and Arithmetic, Tata Institute of Fundamental Research, Bombay 1979

5. Koblitz, N.-p-adic variation of the zeta function of families of varieties defined over finite fields. Comp. Math. 31, pp. 119-218 (1975)

6. Manin, Y.I.-The Hasse-Witt matrix of an algebraic curve. Amer. Math. Soc. Transl. Ser. 45, pp. 245-264, (1965)

7. Manin, Y.I.-On the theory of Abelian varieties over fields of finite characteristics. Izv. Akad. Nauk. SSSR, Ser. Math. 26, pp. 281-292 (1962)

8. Oort, F.-Finite group schemes, local moduli for abelian varieties and lifting problems. Alg. Geom. Oslo 1970, Walters-Noordhoff, 1972

9. Oort, F.-Abelian varieties: Moduli and lifting properties. Alg. Geom. Copenhagen 1978, Lecture Notes in Mathematics, No. 732, Springer Verlag, Berlin, Heidelberg, New York 1979

0. Yui, N.-On the Jacobian varieties of hyperelliptic curves over fields of characteristic p>0. Jour. of Alg. 52, pp. 378-410 (1978)

Department of Mathematics
Princeton University
Fine Hall
Princeton, N.J. 08544

PLATONIC SOLIDS, KLEINIAN SINGULARITIES, AND LIE GROUPS

P. Slodowy
Mathematisches Institut
Universität Bonn
Wegelerstraße 10
D-5300 Bonn
W. Germany

0. Introduction

The aim of this article is to give a survey of a series of mathematical discoveries, from antiquity up to today, which are all related to the classification of the most fundamental, or most simple objects in different mathematical fields. Astonishingly, it turned out that the objects of these different classifications are related to each other by mathematical constructions. However, up to now, these constructions do not explain why the different classifications should be related at all.

In our exposition we shall follow a rough chronological order including some of the speculations in the domain of natural philosophy to which these objects lent themselves by their fundamental character. We also try to keep things as elementary as possible. Thus the article should be understandable to a fairly general audience.

1. Regular Solids

The mathematical definition and classification of regular solids in three-dimensional space is attributed to the Greek mathematician Theaetetus (415-369 B.C.), cf. [65]. There are only five such solids, also called regular polyhedra:

the tetrahedron,

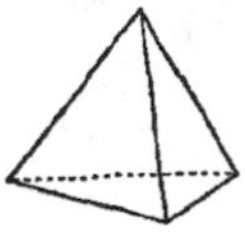

the cube, or hexahedron,

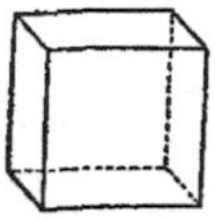

the octahedron,

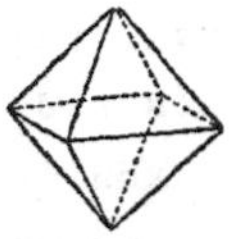

the dodecahedron,

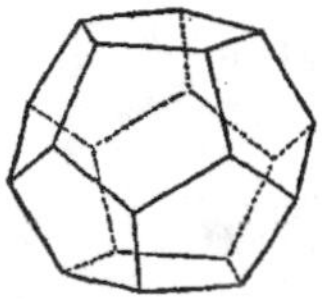

and
the icosahedron

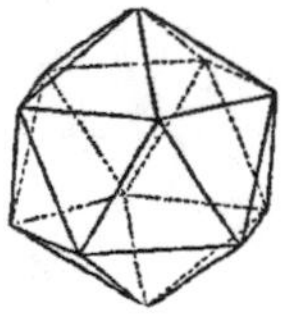

This result is considered to be a most important one in ancient mathematics, and it is believed that the books of Euclid are directed towards the goal of deriving Theaetetus' classification in the final volume XIII.

Let us collect some first data on these solids:

	number of		
solids	faces,	edges,	vertices
tetrahedron	4	6	4
hexahedron	6	12	8
octahedron	8	12	6
dodecahedron	12	30	20
icosahedron	20	30	12

2. Plato's "Timaios"

In the historically first attempt to give an account of the structure of matter in terms of mathematical models, Plato (427-347 B.C.), in his dialogue "Timaios" ([47]), associates the four solids tetrahedron, cube, octahedron, icosahedron to the four "elements" fire, earth, air, water. The dodecahedron is reserved as a symbol for the whole universe Plato considers the four polyhedra attached to the four elements as a kind of "molecules" themselves being composed of two "atomic" planar triangles with angles

$\pi/2,\ \pi/3,\ \pi/6$

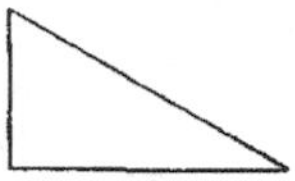

resp.

$\pi/2,\ \pi/4,\ \pi/4$

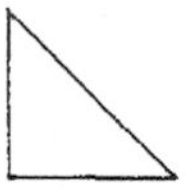

.

The tetrahedron, octahedron, and icosahedron are composed of the first sort of triangles (six of them are formed together to build a face, a regular triangle). Thus Plato obtains "chemical reactions" between the elements fire, air, and water, as for example:

1 unit of fire + 1 unit of water $\rightleftharpoons$ 3 units of air.

3. Kepler's "Mysterium Cosmographicum"

Two thousand years after Plato, in 1596, the young Johannes Kepler published his book "Prodromus dissertationum cosmographicarum, continens mysterium cosmographicum de admirabili proportione orbium coelestium" [32] in which he used the regular polyhedra to describe our planetary system. At that time only six planets were known, Mercury, Venus, Earth, Mars, Jupiter and Saturn, which, according to the model of Copernicus, turned around the sun in circular orbits. Thus each planet determined a sphere around the sun containing its orbit. Kepler remarked that between two successive spheres one could inscribe a conveniently chosen regular polyhedron such that its vertices would lie on the exterior sphere and its faces would touch the interior sphere, in a good approximation at least. Thus one finds

the octahedron	between	Mercury and Venus,
the icosahedron	between	Venus and Earth,
the dodecahedron	between	Earth and Mars,
the tetrahedron	between	Mars and Jupiter
and the cube	between	Jupiter and Saturn

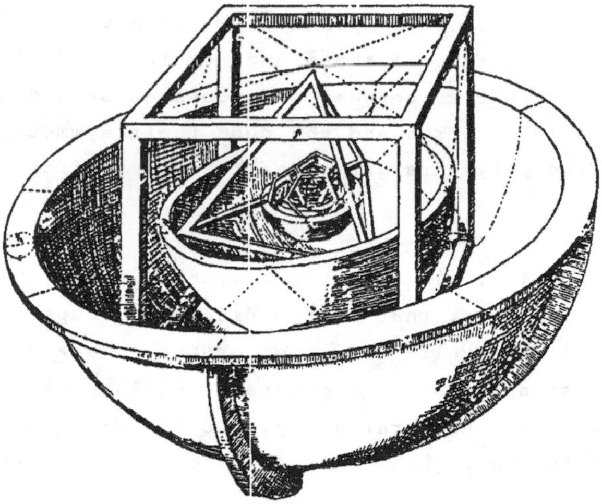

Here is a comparison of the sun-planet distances in Kepler's polyhedral model and in the model of Copernicus. We take the distance sun-earth as a unit.

	Mercury	Venus	Earth	Mars	Jupiter	Saturn
Kepler	0.43	0.76	1	1.44	5.26	9.16
Copernicus	0.39	0.72	1	1.52	5.20	9.55 .

Later Kepler discovered that the orbits were ellipses. Nonetheless, in a second edition of his book in 1621 he still maintained his polyhedral model. Let us add that the seventh planet, Uranus, was not discovered until 1781!

4. Symmetry Groups of Regular Polyhedra

Now let us go to the last century and let us consider the group of rotations $\bar{\Gamma} = \bar{\Gamma}(P)$ transforming a given regular polyhedron P into itself. If we regard P as centered at the origin of euclidean space then we may identify $\bar{\Gamma}$ with a finite subgroup of the group $SO(3,\mathbb{R})$ of all rotations fixing the origin. One easily finds that the axis of rotation of an element $\gamma \in \bar{\Gamma}$ has to pass through either a mid-point of a face, or a mid-point of an edge, or a vertex. We denote the orders of symmetry of these axes p,q,r respectively. We have listed these numbers as well as the structure of the group $\bar{\Gamma}$ in the table below. Because of the geometric duality between the icosahedron and the dodecahedron, resp. the octahedron and the cube their symmetry groups coincide. Thus the five polyhedra give rise to three different finite subgroups of $SO(3,\mathbb{R})$.

To obtain, up to conjugation, all the finite subgroups of $SO(3,\mathbb{R})$ one has to add two infinite series, that of the dihedral groups of order $2n$, $n \geq 2$, and that of the cyclic groups of order n, $n \geq 2$. Following Felix Klein [34] one can associate a degenerate regular polyhedron, a so-called "dihedron", to any dihedral group. Such a dihedron consists of two faces, each of them a regular polygon of order n, attached to each other:

Example $n = 6$:

The corresponding dual polyhedron may be interpreted as an "orange", i.e. a sphere partitioned into n sectors:

Example $n = 6$:

Obviously the rotational symmetry groups of these polyhedra have the structure of a dihedral group. Finally, if one wants, one may consider the cyclic groups as symmetries of certain nonregular solids, i.e., of the pyramids with base a regular polygon:

Example $n = 6$:

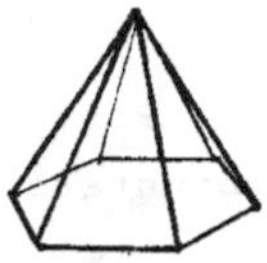

Table:

polyhedron	p	q	r	$\bar{\Gamma}$	$\#\bar{\Gamma}$
pyramid		–		cyclic	n
dihedron	n	2	2	dihedral	2n
orange	2	2	n	dihedral	2n
tetrahedron	3	2	3	$\mathfrak{A}_4$	12
octahedron	3	2	4	$\mathfrak{S}_4$	24
cube	4	2	3	$\mathfrak{S}_4$	24
icosahedron	3	2	5	$\mathfrak{A}_5$	60
dodecahedron	5	2	3	$\mathfrak{A}_5$	60

Here $\mathfrak{S}_m$ (resp. $\mathfrak{A}_m$) denotes the symmetric, (resp. alternating) group of all (resp. of all even) permutations of m letters. For the identification of the groups cf. [34].

Note that the triples p,q,r appearing in the table are exactly the solutions of the diophantine inequality

$$\frac{1}{p} + \frac{1}{q} + \frac{1}{r} > 1$$

with $p,q,r \geq 2$ (see [57], [66] for the explanation).

5. Kleinian Singularities

In his investigations of hypergeometric differential equations with finite monodromy [49], H.A. Schwarz is led to consider the finite subgroups of the projective linear group $PGL(2,\mathbb{C})$ as well as their projective invariants. If one identifies the projective line $\mathbb{P}^1$ with the euclidean sphere S^2 one may regard $PGL(2,\mathbb{C})$ as the group of all conformal transformations of S^2 and $SO(3,\mathbb{R})$ as its maximal compact subgroup of isometries. Thus, up to conjugation, the finite subgroups of $PGL(2,\mathbb{C})$ coincide with those of $SO(3,\mathbb{R})$. The construction of the projective invariants of a finite group $\bar{\Gamma} \subset PGL(2,\mathbb{C})$ reduces to the determination of the so-called relative invariants of the preimage Γ of $\bar{\Gamma}$ in the linear group $SL(2,\mathbb{C})$, which is a double covering of $PGL(2,\mathbb{C})$. Relative invariants of a group are absolute for its commutator subgroup. The absolute invariants of all finite subgroups of $SL(2,\mathbb{C})$ were computed by Felix Klein ([34]) in relation with his investigations of the algebraic equations of degree five. Here is the list of these groups, up to conjugation:

$\mathfrak{C}_n$	the cyclic group of order n
$\mathfrak{D}_n$	the binary dihedral group of order 4n
$\mathfrak{T}$	the binary tetrahedral group of order 24
$\mathfrak{O}$	the binary octahedral group of order 48
$\mathfrak{J}$	the binary icosahedral group of order 120

All these groups, except $\mathfrak{C}_n$ for n odd, are obtained as preimages of corresponding finite subgroups of $SO(3,\mathbb{R}) \subset PGL(2,\mathbb{C})$.

Klein obtained the following results. For each finite group $\Gamma \subset SL(2,\mathbb{C})$ the Γ-invariant polynomials on $\mathbb{C}^2$ are generated by three fundamental invariants X,Y,Z which are subject to a single relation $R(X,Y,Z) = 0$. For conveniently chosen X,Y,Z the relation R takes the following form:

Γ	R
$\mathfrak{C}_n$	$X^n + YZ$
$\mathfrak{D}_n$	$X(Y^2-X^n) + Z^2$
$\mathfrak{T}$	$X^4 + Y^3 + Z^2$
$\mathfrak{O}$	$X^3 + XY^3 + Z^2$
$\mathfrak{J}$	$X^5 + Y^3 + Z^2$

<u>Example</u>: The subgroup $\mathfrak{C}_n \subset SL(2,\mathbb{C})$ is given by the matrices $\begin{pmatrix} \zeta & 0 \\ 0 & \zeta^{-1} \end{pmatrix}$ where ζ runs through all n-th roots of unity. Let us denote coordinates on $\mathbb{C}^2$ by U and V. Then one may take

$$X = UV, \quad Y = U^n, \quad \text{and} \quad Z = -V^n$$

as fundamental invariants satisfying the relation $X^n+YZ = 0$.

The map $q\colon \mathbb{C}^2 \to \mathbb{C}^3$ defined by $q(u,v) = (X(u,v),Y(u,v),Z(u,v))$ is invariant under the action of Γ on $\mathbb{C}^2$. Therefore q factors over the quotient $\mathbb{C}^2 \to \mathbb{C}^2/\Gamma$. On the other hand the polynomials X,Y,Z satisfy the relation $R(X,Y,Z) = 0$ such that q also factors over the imbedding $S \subset \mathbb{C}^3$ where S is the hypersurface $\{(x,y,z) \in \mathbb{C}^3 \mid R(x,y,z) = 0\}$:

$$\begin{array}{ccc} \mathbb{C}^2 & \xrightarrow{\quad q \quad} & \mathbb{C}^3 \\ \downarrow & & \uparrow \\ \mathbb{C}^2/\Gamma & \xrightarrow{\quad \bar{q} \quad} & S \end{array}$$

Furthermore, $\bar{q}\colon \mathbb{C}^2/\Gamma \to S$ is an isomorphism of algebraic varieties.

It is easy to see that the hypersurface S has no singularity except at the origin $0 \in \mathbb{C}^3$. For short, one calls S a Kleinian singularity. Here are some pictures of the real varieties $S \cap \mathbb{R}^3$:

$\mathfrak{C}_2$

$x^2+yz = 0$

$\mathfrak{D}_2$

$x(y^2-x^2)+z^2 = 0$

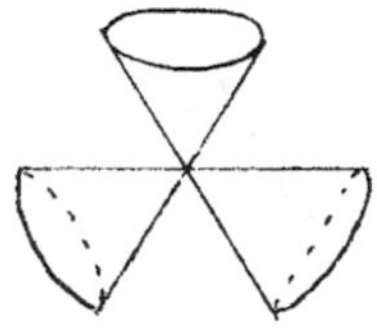

$\mathfrak{T}$

$x(x^2+y^3)+z^2 = 0$

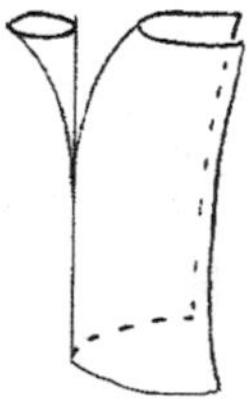

Even before the appearance of Klein's book [34], more or less isolated examples of the singularities introduced above had shown up in the works of algebraic geometers. Let us mention only Schläfli and Cayley, who found some of them as singularities on cubic surfaces ([16],[48]). But from now on Kleinian singularities should be treated and characterized as a class of singularities with specific common properties.

6. The Resolution of Kleinian Singularities

In 1934, P. Du Val characterized the Kleinian singularities as the "isolated singularities of surfaces which do not affect the conditions of adjunction". An essential tool in his investigations [21] is played by the resolution of singularities.

Let S be a complex surface with exactly one singular point $s \in S$. Then a resolution of S is a proper morphism $\pi: \tilde{S} \to S$ (in the category of algebraic or analytic varieties) with the following properties:

- $\tilde{S}$ is a smooth complex surface, i.e. it has no singularities.

- π is a modification of S in s, i.e. the restriction

  $\pi|_{\pi^{-1}(S\setminus s)}: \pi^{-1}(S\setminus s) \to S\setminus s$ is an isomorphism.

A resolution $\pi: \tilde{S} \to S$ of S is called minimal if any other resolution $\pi': \tilde{S}' \to S$ factors over π:

$$\begin{array}{ccc} \tilde{S}' & \dashrightarrow & \tilde{S} \\ & {\scriptstyle \pi'}\searrow \quad \swarrow{\scriptstyle \pi} & \\ & S & \end{array}$$

Minimal resolutions exist and are uniquely determined up to isomorphism ([11], [38]).

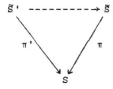

Du Val obtained the following description of the minimal resolution $\pi: \tilde{S} \to S$ of a Kleinian singularity $S = \mathbb{C}^2/\Gamma$. The preimage of the singular point s is a connected union of projective lines

$$\pi^{-1}(s) = C_1 \cup \ldots \cup C_r , \quad C_i \cong \mathbb{P}^1$$

whose self-intersection is -2. This means that the normal bundle of each C_i in $\tilde{S}$ is isomorphic to the cotangent bundle $T^*\mathbb{P}^1$ of the projective line. The pairwise intersections are transverse or void and are symbolized by the dual diagram:
To each curve C_i one associates a vertex $\bullet$, and two vertices are connected by an edge $\bullet\!\!-\!\!\bullet$ if the corresponding curves intersect.
Depending on the finite subgroup $\Gamma \subset SL(2,\mathbb{C})$ one obtains the following dual diagram $\Delta(\Gamma)$

Γ	$\Delta(\Gamma)$	name of $\Delta(\Gamma)$
$\mathfrak{C}_n$	 (n-1 vertices)	A_{n-1}
$\mathfrak{D}_n$	 (n+2 vertices)	D_{n+2}
$\mathfrak{T}$		E_6
$\mathfrak{O}$		E_7
$\mathfrak{J}$	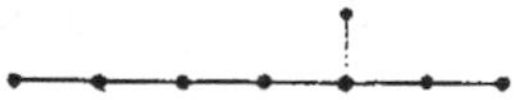	E_8

<u>Examples</u>: In case of the groups $\mathfrak{C}_2$ and $\mathfrak{D}_2$ the real resolution of the real variety $S \cap \mathbb{R}^3$ gives already a quite faithful picture of the complex situation:

$\mathfrak{C}_2$, $S = \{x^2+yz = o\}$ $\tilde{S}$

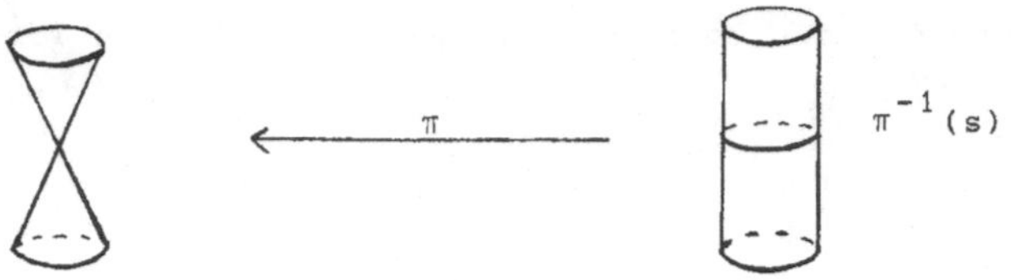

π contracts the circle $(\cong \mathbb{P}^1_{\mathbb{R}})$ to s. The dual diagram $\Delta(\mathfrak{C}_2)$ is of type A_1:

$\mathfrak{D}_2$, $S = \{x(y^2-x^2)+z^2 = 0\}$ $\tilde{S}$

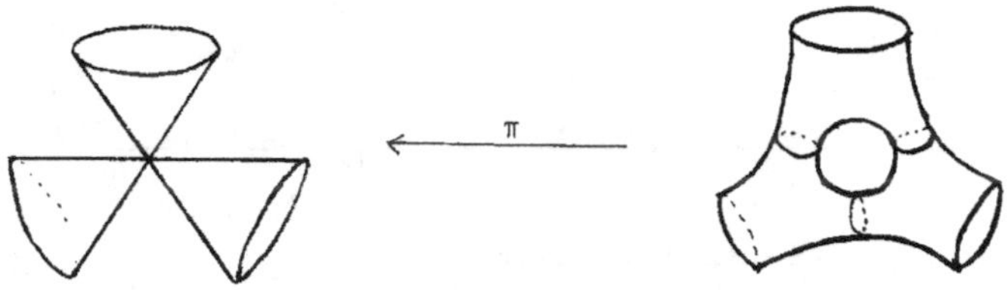

Now $\pi^{-1}(s)$ consists of four circles 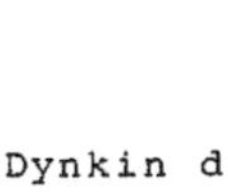

which are contracted to s. The dual diagram $\Delta(\mathfrak{D}_2)$ is of type D_4:

The diagrams $\Delta(\Gamma)$ are exactly the Coxeter or Dynkin diagrams attached to the homogeneous root systems (in which all roots have equal length) in the theory of simple Lie algebras and groups. In the terminology of this theory one may reformulate Du Val's description as follows. The second homology group $H_2(\tilde{S},\mathbb{Z}) \cong \bigoplus_{i=1}^{r} \mathbb{Z}[C_i]$ equipped with the geometric intersection form of cycles on $\tilde{S}$ and the basis $\{[C_1],\ldots,[C_r]\}$ is isomorphic to the root lattice Q of the corresponding root system equipped with the negative of the normalized Killing form and a basis of simple roots. Or, in fewer words, the intersection matrix of the system of curves $C_1,\ldots,C_r$ is the negative of the Cartan matrix of the corresponding root system.

Let us add that a minimal resolution of a Kleinian singularity is easily constructed by iterated blowing up of points. In fact, one may even characterize these singularities as the only double points which can be resolved in such a simple way ([20],[33]). Let us mention also that Du Val had already exploited the relation to root systems to determine, in collaboration with Coxeter, all possible singular points on a degenerate Del Pezzo surface [21].

7. The Fundamental Group of Kleinian Singularities

In the preceding section we have described how one can associate a Dynkin diagram $\Delta = \Delta(\Gamma)$ to a finite subgroup $\Gamma \subset SL(2,\mathbb{C})$. Conversely, one can reconstruct Γ from the diagram.

Let $C = ((c_{ij}))$ be the $r \times r$-Cartan matrix attached to Δ. Then $C = 2I-A$ where I is the identity and A the adjacency matrix of Δ. ($A = ((a_{ij}))$ is defined as follows: $a_{ij} = 0$ if $i = j$ or if the vertices of Δ numbered by i and j are not connected by an edge, otherwise $a_{ij} = 1$). Let $\Pi(C)$ be the group generated by r generators $\gamma_1,\ldots,\gamma_r$ subject to the r relations

$$\gamma_1^{c_{i1}} \cdot \ldots \cdot \gamma_r^{c_{ir}} = 1, \quad i = 1,\ldots,r$$

Then $\Pi(C)$ is isomorphic to Γ. This can be seen in two different ways.

First, using the explicit form of C one easily reduces the above presentation to the following ones:

If Δ is of type A_{n-1}, then $\Pi(C)$ is generated by one element, a, with relation $a^n = 1$.

If Δ is of type D or E, then $\Pi(C)$ is generated by three elements a,b,c with the relations

$$a^p = b^q = c^r = abc,$$

where p,q,r are the degrees of symmetry of the table in section 4. This is a well-known presentation for the binary polyhedral groups, cf. [17].

The second method uses Mumford's description of the fundamental group Π of a neighbourhood boundary of a rational surface singularity in terms of its resolution. Let $C = ((c_{ij}))$ be the intersection matrix for the components of the exceptional divisor of a good resolution of the singularity. Then $\Pi = \Pi(C)$, cf. [27],[44]. In case of a Kleinian singularity $S = \mathbb{C}^2/\Gamma$ a neighbourhood boundary may be identified with the quotient S^3/Γ, where S^3 is a small 3-sphere in $\mathbb{C}^2$ stabl under Γ. Since S^3 is simply connected, and since Γ acts freely on $S^3 \subset \mathbb{C}^2 \setminus 0$ we get $\Pi(C) = \Pi$, where C is the negative of the Cartan matrix of $\Delta(\Gamma)$ by section 6.

By the way, if Γ is the perfect binary icosahedral group then S^3/Γ is Poincaré's well-known example of a manifold with the same integral homology as the sphere S^3 but with different homotopy groups, cf. [15].

8. Deformations and Unfoldings

Starting with Mumford's work [44] there is a rapidly increasing number of investigations studying isolated singularities of algebraic varieties from various points of view. Let us mention as examples the realization of exotic spheres as neighbourhood boundaries ([10]), and the study of the Milnor fibre, its intersection form and monodromy (e.g. [1],[14],

[29],[43],[45]). M. Artin characterizes Kleinian singularities as rational double points ([6]), other characterizations and links to root system notions are established in the work of Brieskorn and Tjurina ([11],[12],[64]).

A central role in most of these investigations is played by certain deformations of the singularities under concern. The general notion of deformation of a singularity is developed at about the same time. It is derived from the deformation theory of complex manifolds (Kodaira-Spencer) which had its origins in Riemann's ideas on moduli spaces for Riemann surfaces.

A deformation of a variety S is a flat morphism $\varphi: X \to (U,u)$ from a variety X (the total space) to a pointed variety (U,u) (the base space) such that the special fibre $\varphi^{-1}(u)$ is isomorphic to S.

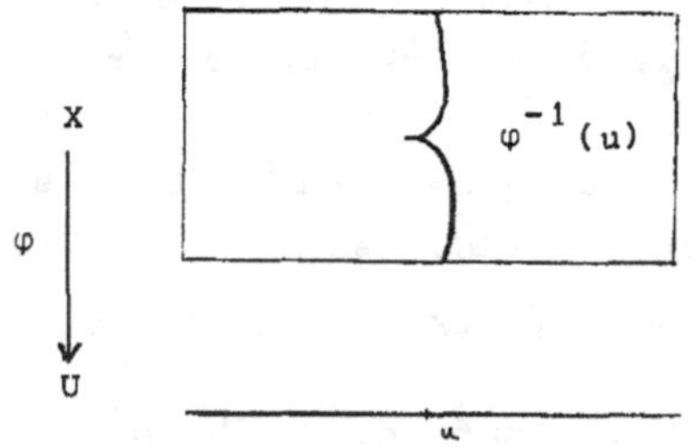

For example, any polynomial function $f: \mathbb{C}^n \to \mathbb{C}$ determines a deformation of any of its fibres $f^{-1}(u)$, $u \in \mathbb{C}$. A deformation $\varphi: X \to (U,u)$ of S is called semiuniversal if any other deformation $\psi: Y \to (V,v)$ of S is induced from φ, more precisely, if, up to isomorphism, ψ can be obtained from φ by a base change $\beta: (V,v) \to (U,u)$ whose differential at v is uniquely determined.

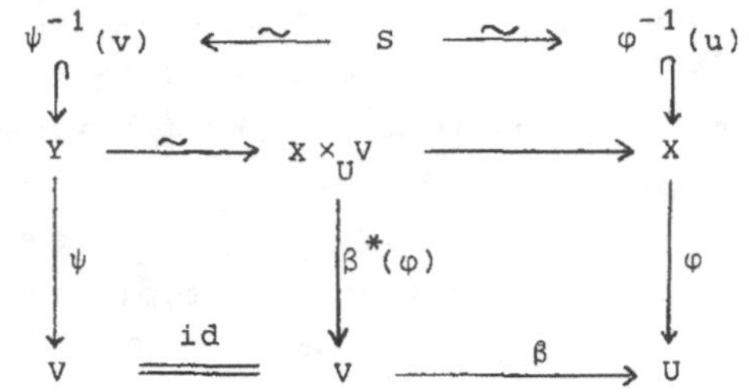

For isolated singularities semiuniversal deformations exist in the category of analytic or henselian space germs and are uniquely determined up to isomorphism ([7],[25],[31],[63]).

Parallel to the developments in algebraic geometry one finds an analogous concentration on the singularities of mappings in the field of differential geometry ([41],[61],[67]). The differential counterpart of the notion of deformation is the notion of unfolding of a function which plays a fundamental role in Thom's theory of morphogenesis ([62]).

An unfolding of a differentiable function $f: \mathbb{R}^n \to \mathbb{R}$ is a differentiable function $F: \mathbb{R}^n \times \mathbb{R}^s \to \mathbb{R}$, for some $s \in \mathbb{N}$, such that the restriction $F_0 = F_{|\mathbb{R}^n \times \{0\}}$ of F to $\mathbb{R}^n \times \{0\}$ coincides with f. The space $\mathbb{R}^s$ is called the parameter space of the unfolding. We may consider F as a differentiable family of functions, i.e. of the restrictions $F_v = F_{|\mathbb{R}^n \times \{v\}}$, $v \in \mathbb{R}^s$. Once we fix a notion of isomorphism for functions (for example, equivalence with respect to coordinate changes on the source $\mathbb{R}^n$ and translations on the image $\mathbb{R}$, which is used by Thom, or contact equivalence) we have a notion of semiuniversal unfolding analogous to the case of deformations. For a function germ $f: (\mathbb{R}^n,0) \to (\mathbb{R},0)$ with algebraically isolated singularity at 0 a semiuniversal unfolding $F: (\mathbb{R}^n \times \mathbb{R}^s,0) \to (\mathbb{R},0)$ exists as a function germ and is uniquely determined up to isomorphism.

Similar definitions and results are also available for complex holomorphic functions. Let $f: (\mathbb{C}^n,0) \to (\mathbb{C},0)$ be a holomorphic germ with an isolated singularity at 0 and $F: (\mathbb{C}^n \times \mathbb{C}^s,0) \to (\mathbb{C},0)$ its semiuniversal unfolding in the sense of Thom. Then the analytic space germ $(S,0) = (f^{-1}(0),0)$ has an isolated singularity at the origin, and the map germ $\varphi: (\mathbb{C}^n \times \mathbb{C}^s,0) \to (\mathbb{C}^1 \times \mathbb{C}^s,0)$ given by $\varphi(x,v) = (F(x,v),v)$ defines a deformation of $(S,0)$ (note $\varphi^{-1}(0) \cong f^{-1}(0)$!) which is versal, i.e. which is the product of a semiuniversal deformation by a trivial factor (whose dimension equals the difference of the so-called right- and contact-codimension of f). This trivial factor disappears when f is quasihomogeneous in some coordinate system, for example, when f is a polynomial function $R: \mathbb{C}^3 \to \mathbb{C}$ of the table in section 5 and when $S = R^{-1}(0)$ is a Kleinian singularity. For simplicity, let us assume this case now.

An important subset of $U = \mathbb{C}^1 \times \mathbb{C}^s$ is the discriminant $D(\varphi)$, i.e. the set of critical values of φ. Its local structure near some point $u \in D(\varphi)$ contains much information about the singularities in the fiber $\varphi^{-1}(u)$. The critical set $K_{\mathbb{C}}$ of the projection of $D(\varphi) \subset \mathbb{C}^1 \times \mathbb{C}^s$ onto $\mathbb{C}^s$ coincides with the image of the singularities of $D(\varphi)$ under this projection. It is called the (complex) catastrophe set and it decomposes into two subsets $K_{\mathbb{C}} = B_{\mathbb{C}} \cup M_{\mathbb{C}}$, where $B_{\mathbb{C}} = \{v \in \mathbb{C}^r \mid F_v$ has a degenerate critical point$\}$ is the bifurcation set and where $M_{\mathbb{C}} = \{v \in \mathbb{C}^r \mid F_v$ has the same value in at least two different critical points$\}$ is the Maxwell set ([62]).

<u>Example</u>: Let $f = R: \mathbb{C}^3 \to \mathbb{C}$, $R(x,y,z) = x^4+yz$. Then F is given by $F: \mathbb{C}^3 \times \mathbb{C}^2 \to \mathbb{C}$, $F(x,y,z,u,v) = x^4+yz+ux^2+vx$. The discriminant of φ in $\mathbb{C}^3$ looks like the "queue d'aronde"

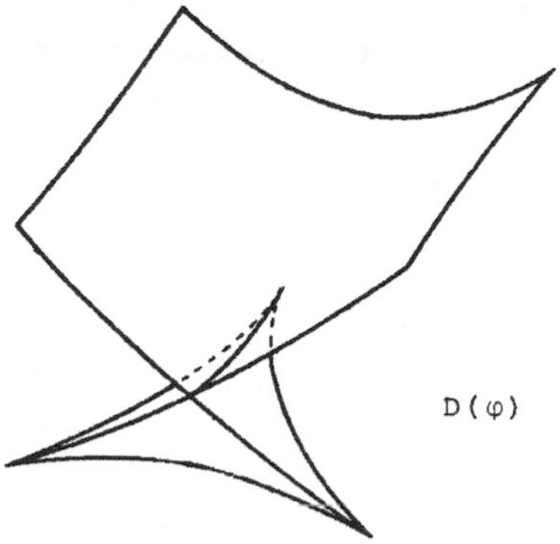

$D(\varphi)$

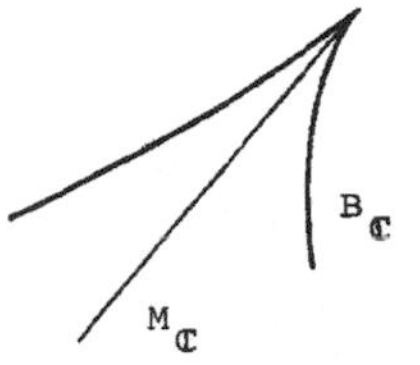

and the catastrophe set $K_{\mathbb{C}}$ in $\mathbb{C}^2$ consists of a cusp $B_{\mathbb{C}}$ and a line $M_{\mathbb{C}}$.

9. Elementary Catastrophes and Simple Singularities

In Thom's catastrophe theory [62] the universal unfolding $F: \mathbb{R}^n \times \mathbb{R}^r \to \mathbb{R}$ of a function $f: \mathbb{R}^n \to \mathbb{R}$ is interpreted as a stable r-parameter family of gradient dynamics on $\mathbb{R}^n$:

$$X_u(x) = -(\mathrm{grad}_{\mathbb{R}^n} F_u)(x), \quad x \in \mathbb{R}^n, \; u \in \mathbb{R}^r.$$

If one assumes that F describes a natural process extended over the substrate space $\mathbb{R}^r$ whose local state in a point $u \in \mathbb{R}^r$ is given by an equilibrium position of the "internal" dynamic X_u, then the (real) catastrophe set $K = B \cup M$, $B = \{u \in \mathbb{R}^r \mid F_u$ has a degenerate critical point$\}$, $M = \{u \in \mathbb{R}^r \mid F_u$ has the same value in at least two critical points$\}$, describes the points $u \in \mathbb{R}^r$ where the structural or statistical stability of an equilibrium of X_u breaks down. Thus one should perceive discontinuous changes of local states along (at least certain strata of) the catastrophe set.

The classification of the so-called elementary catastrophes is essentially the determination of all function germs $f: (\mathbb{R}^n,0) \to (\mathbb{R},0)$ whose universal unfolding can be realized by a parameter space of dimension $r \leq 4$, i.e. which give rise to stable families of gradient dynamics on space-time. Though there are infinitely many functions f with this property (up to equivalence) the resulting catastrophe sets belong to a set of only seven different types (resp. eight if one accounts for the empty set $\emptyset$).

name of catastrophe	$f: \mathbb{R}^n \to \mathbb{R}$, $n \geq 1$ (resp.2)	unfolding dimension r	type of $f_{\mathbb{C}}: \mathbb{C}^3 \to \mathbb{C}$
∅	$\pm x_1^2 \pm x_2^2 \pm \ldots \pm x_n^2$	0	A_1
fold	$x_1^3 \pm x_2^2 \pm \ldots \pm x_n^2$	1	A_2
cusp	$\pm x_1^4 \pm x_2^2 \pm \ldots \pm x_n^2$	2	A_3
swallowtail	$x_1^5 \pm x_2^2 \pm \ldots \pm x_n^2$	3	A_4
butterfly	$\pm x_1^6 \pm x_2^2 \pm \ldots \pm x_n^2$	4	A_5
elliptic umbilic	$x_1^3 + x_1 x_2^2 \pm x_3^2 \pm \ldots \pm x_n^2$	3	D_4
hyperbolic umbilic	$x_1^3 - x_1 x_2^2 \pm x_3^2 \pm \ldots \pm x_n^2$	3	D_4
parabolic umbilic	$\pm x_1^4 + x_1 x_2^2 \pm x_3^2 \pm \ldots \pm x_n^2$	4	D_5

In the last column we have considered the complexification $f_{\mathbb{C}}: \mathbb{C}^3 \to \mathbb{C}$ of the respective function $f: \mathbb{R}^3 \to \mathbb{R}$ in three real variables. These polynomials occurred in section 5 as the relation R between the fundamental invariants of certain finite subgroups of $SL(2,\mathbb{C})$, namely, the cyclic groups of order ≤ 6, and the binary dihedral groups $\mathfrak{D}_2$, $\mathfrak{D}_3$. The Dynkin diagram of the resolution of the corresponding Kleinian singularity, given by $f_{\mathbb{C}} = R = 0$, is given in that last column.

The condition on a function germ to have a universal unfolding with parameter space of dimension ≤ 4 is not intrinsic from a mathematical point of view. A natural condition on the unfolding to single out all "Kleinian" functions $R: \mathbb{C}^3 \to \mathbb{C}$ up to the addition of squares in additional variables was found by Arnol'd [3]. He calls a germ f <u>simple</u> if the perturbed functions F_u in its semiuniversal unfolding F present only a finite number of non-equivalent singular points. He showed that up to addition or deletion of squares in "dummy" variables it suffices to look at the simple germs $f: (\mathbb{C}^3,0) \to (\mathbb{C},0)$ and that these are given exactly by the Kleinian relations $R: \mathbb{C}^3 \to \mathbb{C}$

tabulated in section 5. For real germs, in addition, one only has to take care of signs, similarly as in the list above. Also, if one formulates Arnol'd's simplicity condition for deformations of hypersurfaces $f = 0$, one gets the analogous result.

10. Simple Lie Groups and Kleinian Singularities

A complex Lie group G is a complex manifold equipped with an analytic group structure. It is called (almost) simple if it contains no normal subgroup of positive dimension. Here we cannot give an account of the elaborate structure theory of these groups. We merely recall certain facts ([9],[28],[58]).

Any simple complex Lie group G is a finite central quotient of its universal covering $\tilde{G}$ whose center is finite. These universal coverings, i.e. the simply connected simple groups are classified by their Dynkin diagrams which, in a sense, encode all relevant information on the group.

Dynkin diagram Δ

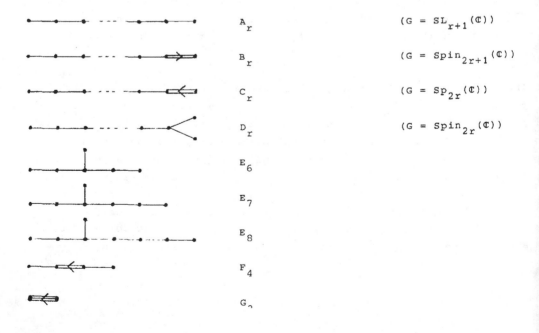

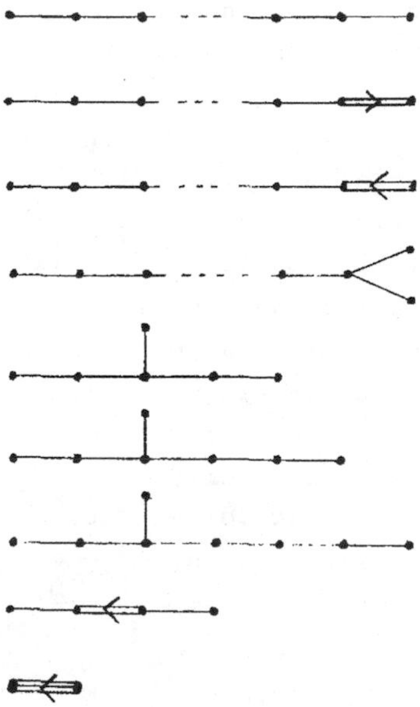

A_r $(G = SL_{r+1}(\mathbb{C}))$

B_r $(G = Spin_{2r+1}(\mathbb{C}))$

C_r $(G = Sp_{2r}(\mathbb{C}))$

D_r $(G = Spin_{2r}(\mathbb{C}))$

E_6

E_7

E_8

F_4

G_2

The number r of nodes in the diagram Δ is called the rank of G. It coincides with the dimension of a maximal torus $T \subset G$.

A group G of rank r has r fundamental irreducible representations

$$\rho_i: G \to GL(V_i), \quad i = 1,\dots,r$$

on finite-dimensional complex vector spaces V_i. To each representation ρ_i is associated the corresponding character

$$\chi_i: G \to \mathbb{C}$$

$$\chi_i(g) = \text{trace } \rho_i(g).$$

The map $\chi: G \to \mathbb{C}^r$ defined by

$$\chi(g) = (\chi_1(g),\dots,\chi_r(g))$$

is invariant under conjugation

$$\chi(xgx^{-1}) = \chi(g), \quad g,x \in G$$

since the trace is. Thus each fibre of χ is a union of conjugacy classes of G. In fact one can show that χ is the quotient of G by its conjugation action in the category of algebraic varieties [58].

<u>Example</u>: Let us consider the special linear group $SL_n(\mathbb{C})$ of rank $r = n-1$. As a maximal torus T we may choose the diagonal matrices. The r fundamental representations $\rho_i: G \to GL(V_i)$ are afforded by the exterior powers

$$V_i = \Lambda^i \mathbb{C}^n, \quad i = 1,\dots,n-1$$

and the corresponding characters coincide, up to sign, with the non-trivia coefficients of the characteristic polynomial:

$$\begin{aligned} \text{char}(g) &= \det(\lambda - g) \\ &= \lambda^n - \text{trace}(g)\lambda^{n-1} + \text{trace}(\Lambda^2 g)\lambda^{n-2} - \dots + (-1)^n. \end{aligned}$$

Thus χ may be regarded as associating to $g \in SL_n(\mathbb{C})$ its characteristic polynomial char(g). A fibre of χ now consists of all matrices in $SL_n(\mathbb{C})$ having the same characteristic polynomial, i.e. the same eigenvalues (taking into account multiplicities). In particular, by the theory of Jordan normal forms one sees that each fibre consists of a finite number of conjugacy classes.

Also for general G one has the result that all fibres of χ consist of a finite union of conjugacy classes. Furthermore, χ is a flat morphism. Thus all fibres have the same dimension $d = \dim G - r$. The special fibre $\chi^{-1}(\chi(e))$ containing the neutral element e is called the unipotent variety $\mathrm{Uni}(G)$ of G. Its elements are transformed into unipotent matrices under any rational representation of G.

The detailed investigation of the morphism χ is due to Kostant and Steinberg ([37],[58]). With the knowledge of their works and those of Brieskorn on simultaneous resolution for Kleinian singularities Grothendieck conjectured, and Brieskorn later established a profound relationship between Kleinian singularities and the corresponding Lie groups of type A_r, D_r, E_r:

The unipotent variety $\mathrm{Uni}(G)$ is the closure of its unique conjugacy class of maximal dimension d. This class is called regular and its complement in $\mathrm{Uni}(G)$ has codimension 2, i.e. dimension $d-2$. And again, this complement is the closure of a single conjugacy class, called subregular class. Let $X \subset G$ be a slice of dimension $r+2 = \dim G - d + 2$ transversal to the subregular class at an element x. Then Grothendieck conjectured

<u>Theorem (Brieskorn [13]):</u> Let G be a simply connected Lie group of type A_r, D_r, or E_r. Then

i) The intersection $X \cap \mathrm{Uni}(G)$ is a Kleinian singularity of the same type as G.

ii) The restriction $\chi_{|X}: X \to \mathbb{C}^r$ of χ to X realizes a semiuniversal deformation of $X \cap \mathrm{Uni}(G)$.

<u>Example:</u> In the case $G = SL_n(\mathbb{C})$ all regular (resp. subregular)

unipotent elements are conjugate to the matrix

$$\begin{pmatrix} 1 & 1 & & & 0 \\ & 1 & \cdot & & \\ & & \cdot & \cdot & \\ & & & \cdot & 1 \\ 0 & & & & 1 \end{pmatrix} \qquad \text{(resp.} \quad \begin{pmatrix} 1 & 0 & & & & 0 \\ & 1 & 1 & & & \\ & & \cdot & \cdot & & \\ & & & \cdot & \cdot & \\ & & & & \cdot & 1 \\ 0 & & & & & 1 \end{pmatrix} \quad).$$

Using the logarithm from SL_n to its Lie algebra sL_n one can easily construct a slice X and verify the theorem by computation (cf. [2]). In the simplest case $n = 2$ one can even take $X = SL_2(\mathbb{C})$. Then χ is given by

$$\chi\begin{pmatrix} a & b \\ c & d \end{pmatrix} = a+d$$

and

$$\chi^{-1}(\chi(e)) = \{\begin{pmatrix} 1+x & y \\ z & 1+u \end{pmatrix} | x+u = 0, \quad xu-yz = 0\}$$

$$= \{(x,y,z) \in \mathbb{C}^3 | x^2+yz = 0\}.$$

Combining the theorem above with other constructions in Lie group theory one obtains much information about Kleinian singularities and their deformation theory, e.g. simultaneous resolution, neighbouring singularities, discriminant locus, and more. Here we only want to sketch how one finds the minimal resolution of a Kleinian singularity in the Lie theoretic context. For details on this and the other topics we refer to [58],[52],[53].

A Lie subgroup $P \subset G$ is called parabolic if the quotient space G/P is a compact manifold. In that case it is in fact a projective variety. Parabolic subgroups of minimal dimension are called Borel subgroups. They are all conjugate in G, and they may be characterized as the maximal solvable subgroups of G. The normalizer $N_G(P)$ of a parabolic subgroup P in G coincides with P. Therefore the set of all conjugates of P may be identified with the variety G/P.

<u>Example:</u> In $G = SL_n(\mathbb{C})$ the parabolic subgroups are exactly the stabilizers of flags $0 \subset V_{i_1} \subset \dots \subset V_{i_k} \subset \mathbb{C}^n$, $\dim V_{i_j} = i_j$. Thus the homogeneous spaces G/P correspond to the various flag varieties. The Borel subgroups occur as the stabilizers of maximal flags

$$0 \subset V_1 \subset \dots \subset V_{n-1} \subset \mathbb{C}^n.$$

In particular, all Borel subgroups in $SL_n(\mathbb{C})$ are conjugate to the subgroup consisting of upper triangular matrices.

Let $\mathfrak{B}$ denote the variety of all Borel subgroups of G and consider the incidence variety

$$I = \{(x,B) \in \mathrm{Uni}(G) \times \mathfrak{B} \mid x \in B\}$$

with the natural projections

$$\begin{array}{ccc} & I & \\ {}^{p}\swarrow & & \searrow^{q} \\ \mathrm{Uni}(G) & & \mathfrak{B} \end{array}$$

Then q identifies I with the cotangent bundle $T^*\mathfrak{B}$ of $\mathfrak{B}$ and, by a result of T.A. Springer, p is a G-equivariant resolution of the singularities of $\mathrm{Uni}(G)$. Now let G be as in the theorem, $x \in \mathrm{Uni}(G)$ a subregular element and X a transversal slice at x. Then $S = X \cap \mathrm{Uni}(G)$ is a Kleinian singularity. By the G-equivariance of p one sees that the restriction

$$p: \tilde{S} = p^{-1}(S) \to S$$

of the resolution $I \to \mathrm{Uni}(G)$ over S is again a resolution, in fact, by results of Steinberg, Tits and Esnault it is the minimal resolution of S. The exceptional set $p^{-1}(x)$ has the form

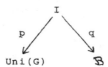

$$p^{-1}(x) = \{(x,B) \in \{x\} \times \mathfrak{B} \mid x \in B\}$$

$$\cong \quad \mathfrak{B}_x = \{B \in \mathfrak{B} \mid x \in B\}.$$

Thus $\mathfrak{B}_x$ must consist of a union of projective lines intersecting in a configuration as prescribed by the Dynkin diagram Δ of G.

After the choice of a maximal torus T_0 and a Borel subgroup $B_0 \supset T$ one may identify the set $|\Delta|$ of nodes of Δ with the set of simple roots. Here we shall regard $|\Delta|$ as the set of conjugacy classes of minimal proper parabolic subgroups, i.e. which are not Borel subgroups. A representative P_α of each class $\alpha \in |\Delta|$ is given by the subgroup of G generated by B_0 and the root subgroup $U_{-\alpha}$ corresponding to the negative of the simple root α. Let $\mathfrak{P}_\alpha$ denote the variety G/P_α of all subgroups conjugate to P_α. Then there is a natural map

$$f_\alpha : \mathfrak{B} \to \mathfrak{P}_\alpha$$

which we may identify with $G/B_0 \to G/P_\alpha$, and which sends a Borel subgroup $B \in \mathfrak{B}$ to the unique parabolic $P \in \mathfrak{P}_\alpha$ containing B. Thus each fibre $f_\alpha^{-1}(P)$ consists of the Borel subgroups contained in P. Since P has semisimple rank 1 this is a projective line. We call any fibre of f_α a line of type α in $\mathfrak{B}$. Then Steinberg and Tits showed that $\mathfrak{B}_x$ is a union of lines of types $\alpha \in |\Delta|$, each type occurring exactly once (G of type A,D,E) and intersecting each other as prescribed by the bonds of Δ.

<u>Example</u>: Let $G = SL_n(\mathbb{C})$. Then the minimal proper parabolic subgroups are the stabilizers of almost maximal flags

$$0 \subset V_1 \subset \ldots \subset V_{i-1} \subset V_{i+1} \subset \ldots \subset V_{n-1} \subset \mathbb{C}^n$$

where the missing dimension i may run from 1 to $n-1$ in accordance with the Dynkin diagram Δ of type A_{n-1}:

The map $f_i : \mathcal{B} \to \mathcal{F}_i$ corresponds to $(V_1 \subset \ldots \subset V_i \subset \ldots \subset V_{n-1}) \mapsto (V_1 \subset \ldots \subset V_{i-1} \subset V_{i+1} \subset \ldots \subset V_{n-1})$ on the flag level, and its fibres are immediately recognized as projective lines. We leave it to the reader as an exercise to write down $\mathcal{B}_x$ for the matrix

$$x = \begin{pmatrix} 1 & 0 & & & 0 \\ 1 & 1 & & & \\ & 1 & \ddots & & \\ & & \ddots & \ddots & \\ 0 & & & 1 & 1 \end{pmatrix} \qquad \text{(cf. [52],[59]).}$$

11. Representations and Dynkin Diagrams

In section 6 we have attached a Dynkin diagram Δ to any finite subgroup $\Gamma \subset SL_2(\mathbb{C})$ by resolving the Kleinian singularity $S = \mathbb{C}^2/\Gamma$. A different way to obtain this diagram, i.e. without involving S, was recently discovered by McKay who established a relation between the irreducible representations of Γ and the extended Dynkin diagram $\tilde{\Delta}$ of Δ (which determines Δ uniquely).

type of Δ	$\tilde{\Delta}$
A_r	
D_r	
E_6	
E_7	
E_8	

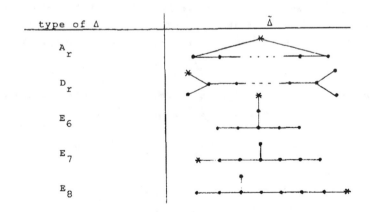

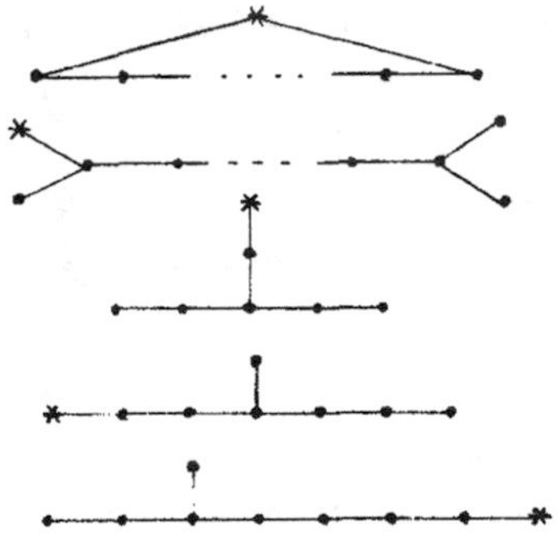

(Here a $*$ denotes the extra node one has to add to Δ).

Let us quickly recall some elementary facts on the representation theory of a finite group Γ (cf. [51]):

- Any representation of Γ on a finite-dimensional complex vector space is completely reducible
- The number of irreducible complex representations (up to equivalence) equals the number of conjugacy classes of Γ.
- Let $R_0 = \text{triv.}$, $R_1,\dots,R_r$ denote representatives of the irreducible representations of Γ and let $d_i = \dim R_i$ denote their degree. Then

$$\sum_{i=0}^{r} d_i^2 = \text{card}(\Gamma).$$

In the following, when we speak of a representation of Γ we shall silently understand by that its equivalence class, i.e. its character. Now fix a finite subgroup $\Gamma \subset SL_2(\mathbb{C})$ and let N denote the corresponding two-dimensional representation (which is irreducible except for cyclic Γ). For each $i = 0,\dots,r$ we can decompose the tensor product

$$N \otimes R_i = \bigoplus_{j=0}^{r} a_{ij} R_j$$

where a_{ij} denotes the multiplicity of R_j in $N \otimes R_i$. Let $A = ((a_{ij}))$ be the corresponding matrix and let $I \in M_{r+1}(\mathbb{Z})$ be the unit matrix. Then McKay observed:

> The matrix $2I-A$ is the Cartan matrix of the extended Dynkin diagram $\tilde{\Delta}$ associated to Γ.

This observation was based first on an explicit verification ([42]). A more systematic derivation was given afterwards by Steinberg ([60], cf. also [26]).

McKay's results looks nicer if we interpret it in terms of the diagram $\tilde{\Delta}$. Then to each node of $\tilde{\Delta}$ corresponds an irreducible representation. Up to symmetry of $\tilde{\Delta}$ the trivial representation R_0 belongs to the extra node $*$. If we tensor a representation R_i with N then the product $N \otimes R_i$ decomposes into the direct sum of the representations R_j whose nodes are connected to that of R_i.

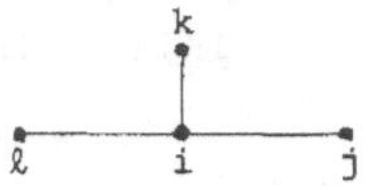

$$N \otimes R_i = R_j \oplus R_k \oplus R_\ell .$$

Another interpretation of McKay's result is in terms of eigenvectors and eigenvalues of the Cartan matrix of $\tilde{\Delta}$. Let χ (resp. χ_j) denote the character $\Gamma \to \mathbb{C}$, $\gamma \mapsto$ trace $\rho(\gamma)$, attached to the representation ρ of Γ on N (resp. R_j), and let $\gamma_0 = 1, \gamma_1, \ldots, \gamma_r$ denote representatives of the conjugacy classes in Γ. Then the characters χ_j are determined by their values $\chi_j(\gamma_k)$ on the γ_k. The $(r+1)\times(r+1)$-matrix constituted by these values is called the character table of Γ:

	γ_0	γ_1	...	γ_k	...	γ_r
R_0	1	1	...	1	...	1
R_1	d_1					
⋮						
R_j	d_j			$\chi_j(\gamma_k)$		
⋮						
R_r	d_r					

The formula $N \otimes R_i = \bigoplus_{j=0}^{r} a_{ij} R_j$ now implies

$$\chi(\gamma_k)\chi_i(\gamma_k) = \sum_{j=0}^{r} a_{ij}\chi_j(\gamma_k)$$

or

The k-th column of the character table is an eigenvector of A (thus of $2I-A$) for the eigenvalue $\chi(\gamma_k)$ (resp. $2-\chi(\gamma_k)$).

In particular, we see that the first column $(d_0, d_1, \ldots, d_r)$ of the degrees is annihilated by $2I-A$. This corresponds to the fact that the numbers $d_1, \ldots, d_r$ are the coefficients of the highest root in the root system associated to Δ. On the side of the singularity $S = \mathbb{C}^2/\Gamma$ the highest root represents the fundamental cycle in $H_2(\tilde{S}, \mathbb{Z})$ where $\tilde{S}$ is the minimal resolution of S. Thus the numbers $d_1, \ldots, d_r$ coincide with the multiplicities of the components of the exceptional divisor in $\tilde{S}$ (cf. [46]).

12. Representations and Resolutions

McKay's observation naturally raises the question whether there is any direct relation between the (non-trivial) irreducible representations of Γ and the irreducible components of the exceptional divisor $\pi^{-1}(0)$ in the minimal resolution $\pi: \tilde{S} \to S$ of the Kleinian singularity $S = \mathbb{C}^2/\Gamma$. In 1981 such a relation was found by Gonzalez-Sprinberg and Verdier (cf. [23],[24], and [35] for an independent proof).

Their basic idea is as follows. Since Γ acts freely on $\dot{\mathbb{C}}^2 = \mathbb{C}^2 \setminus 0$ the quotient map $\dot{\mathbb{C}}^2 \to \dot{\mathbb{C}}^2/\Gamma = S \setminus 0$ is a principal Γ-bundle. Thus to each representation R of Γ we can assign the associated vector bundle

$$\dot{V}(R) = \dot{\mathbb{C}}^2 \times^{\Gamma} R.$$

Since $S \setminus 0$ is isomorphic to $\tilde{S} \setminus \pi^{-1}(0)$ we obtain a bundle on the complement of the exceptional divisor on $\tilde{S}$. This bundle $\dot{V}(R)$ can be extended in a natural way to a bundle $V(R)$ on the whole of $\tilde{S}$. For

$R = R_i$, a non-trivial irreducible representation, one proves that the first Chern class $c_1(V(R_i)) \in H^2(\tilde{S},\mathbb{Z})$ is dual to the class $[C_i] \in H_2(\tilde{S},\mathbb{Z})$ of the exceptional component C_i corresponding to R_i in the Dynkin diagram Δ. Moreover, the correspondence

$$R \mapsto V(R)$$

extends to an additive (not multiplicative!) isomorphism of the representation ring $R(\Gamma)$ of Γ to the Grothendieck ring $K(\tilde{S})$ of vector bundles on $\tilde{S}$.

If we identify $H_2(S,\mathbb{Z})$ with the root lattice Q corresponding to Δ and $[C_i]$ with a simple root α_i then $H^2(\tilde{S},\mathbb{Z})$ may be identified with the weight lattice P

$$P = \bigoplus_{i=1}^{r} \mathbb{Z}\, w_i$$

generated by fundamental dominant weights w_i. We then have $c_1(V(R_i)) = w_i$.

This formulation leads to interesting questions which are related to the still unsolved problem of a satisfactory understanding of the relation between a binary polyhedral group Γ and its corresponding Lie group G.

In section 10 we have seen how the resolution $\pi\colon \tilde{S} \to S$ is realized inside the resolution $T^*\mathfrak{B} \to \mathrm{Uni}(G)$ of the unipotent variety of G. From the description of $\mathfrak{B}_x \cong \pi^{-1}(0)$ and the theory of Schubert cycles on $\mathfrak{B}$ (cf. [8],[18]) it follows that the inclusion $\iota\colon \tilde{S} \hookrightarrow T^*\mathfrak{B}$ induces isomorphisms

$$\iota_*\colon H_2(\tilde{S},\mathbb{Z}) \xrightarrow{\sim} H_2(T^*\mathfrak{B},\mathbb{Z}) = H_2(\mathfrak{B},\mathbb{Z})$$

$$\iota^*\colon H^2(\tilde{S},\mathbb{Z}) \xleftarrow{\sim} H^2(T^*\mathfrak{B},\mathbb{Z}) = H^2(\mathfrak{B},\mathbb{Z}) .$$

To each fundamental dominant weight w_i is associated a homogeneous line bundle

$$\mathcal{L}_i = G\times^B \mathbb{C}$$

on $\mathfrak{B} = G/B$, where B acts linearly on $\mathbb{C}$ by the character $-w_i$. Under the identification of P with $H^2(\tilde{S},\mathbb{Z})$ and $H^2(\mathfrak{B},\mathbb{Z})$ we then have

$$c_1(\mathcal{L}_i) = w_i.$$

Let L_i be the pull back of $\mathcal{L}_i$ to $T^*\mathfrak{B}$. Then, because of the rationality of S we must have an isomorphism

$$L_i|_{\tilde{S}} \cong \Lambda^{\dim R_i}(V(R_i))$$

where $\Lambda^{\dim R_i}(V(R_i))$ denotes the determinant bundle of $V(R_i)$. One question is the following: Is there a "natural" vector bundle V_i on $T^*\mathfrak{B}$ of dimension $\dim R_i$ (possibly the pull back of a homogeneous bundle on $\mathfrak{B}$?) such that

$$V_i|_{\tilde{S}} \cong V(R_i) \quad ?$$

Let us add that the fundamental irreducible representations ρ_i of G are afforded by the spaces $H^0(\mathfrak{B}, \mathcal{L}_i)$ of global sections of the $\mathcal{L}_i$. Thus another question: Is there a deeper relation between the non-trivial irreducible representations of Γ and the fundamental irreducible representations of G?

13. Further Developments

Besides the homogeneous Dynkin diagrams of type A,D,E there are four other types of Dynkin diagrams, the series B_n and C_n, and F_4 and G_2 which correspond to root systems with roots of different lengths. In some sense these diagrams are quotients of A,D,E-diagrams by certain diagram symmetries. Most of the connections between singularities, Lie groups, and representations of binary polyhedral groups go through for diagrams of type B,C,F,G once the right notion of symmetry has been found for the objects involved. The details for this may be found in [52],[53],[55], and [5],[19]. In [52] one also finds an extension of the results of section 10 to algebraic groups over arbitrary fields (not necessarily algebraically closed, cf. [52]App.1) subject to a mild restriction on the characteristic. Modifying the approach of [13],[52] by considering the conjugation representation on infinitesimally symmetric spaces the authors of [50],[50'] obtain deformations of simple singularities of dimensions different from two, e.g. of the simple curve singularities.

A different way to obtain the Dynkin diagram of a simple singularity is to look at the intersection form of its Milnor fibre. When $f: \mathbb{R}^2 \to \mathbb{R}$ is any isolated singularity function, then A'Campo [1] and Husein-Zade [29] showed how to obtain this form by a conveniently chosen unfolding of f. In most cases their method provides an immediate realization of the Dynkin diagram.

<u>Example</u>: $f: \mathbb{R}^2 \to \mathbb{R}$ given by x^5+y^3 (type E_8). The zero set of a nice "morsification" $\tilde{f}$ looks as follows

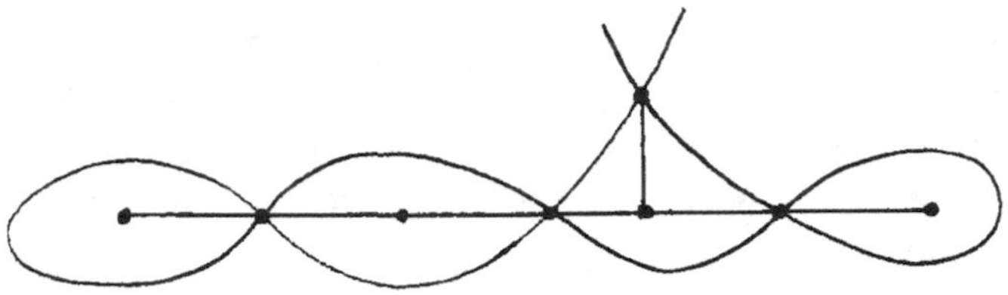

The diagram is obtained by simply connecting the critical points of $\tilde{f}$.

The intersection form of the Milnor fibre is also the clue to relating more general singularities to more general Dynkin diagrams. For example the deformation theory of the simply elliptic and cusp singularities is related to the theory of conjugacy classes in certain groups attached to Kac-Moody algebras with diagrams of the form $T_{p,q,r}$,

$\frac{1}{p} + \frac{1}{q} + \frac{1}{r} \leq 1$:

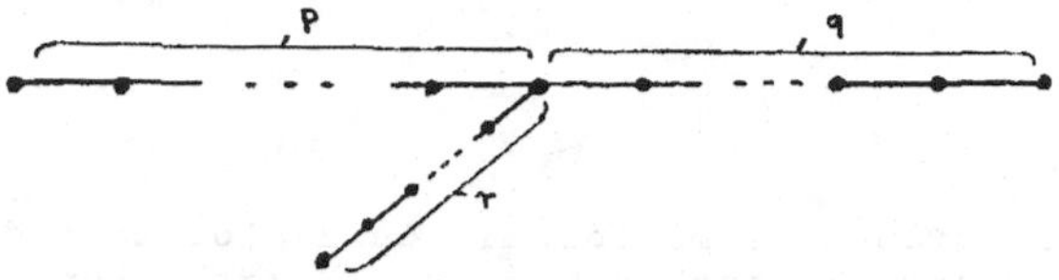

For details cf. [39],[40],[54],[56].

Finally, let us mention that Dynkin diagrams of type A,D,E (resp. their extensions) appear in the classification of quivers of finite representation type (Gabriel [22], for connections of quivers to generalized root systems see [30]) and in the classification of singular fibres in elliptic pencils (Kodaira [36]).

References

[1] A'Campo, N.: Le groupe de monodromie du deploiement des singularités isoleés de courbes planes I; Math. Ann. 213, 1-32 (1975), II, Proc. Int. Cong. Math. Vancouver, Vol I, 395-404 (1974)

[2] Arnol'd, V.I.: On matrices depending on parameters; Russian Math. Surveys 26, 29-43 (1971)

[3] Arnol'd, V.I.: Normal forms for functions near degenerate critical points, the Weyl groups of A_k, D_k, E_k and Lagrangian singularities; Functional Anal. Appl. 6, 254-272 (1972)

[4] Arnol'd, V.I.: Critical points of smooth functions; Proc. Int. Cong. Math. Vancouver, Vol I, 19-39 (1974)

[5] Arnol'd, V.I.: Critical points of functions on a manifold with boundary, the simple Lie groups B_k, C_k, F_4 and singularities of evolutes; Russian Math. Surveys 33, 5, 99-116 (1978)

[6] Artin, M.: On isolated rational singularities of surfaces; Amer. J. Math. 88, 129-136 (1966)

[7] Artin, M.: Lectures on deformations of singularities; Tata Institute, Bombay, 1976

[8] Bernstein, I.N., Gel'fand, I.M., Gel'fand, S.I.: Schubert cells and the cohomology of the spaces G/P; Russian Math. Surveys 28, 3, 1-26 (1973)

[9] Borel, A.: Linear Algebraic Groups; Benjamin, New York 1969

[10] Brieskorn, E.: Beispiele zur Differentialtopologie von Singularitäten; Inventiones math. 2, 1-14 (1966)

[11] Brieskorn, E.: Über die Auflösung gewisser Singularitäten von holomorphen Abbildungen, Math. Ann. 166, 76-102 (1966)

[12] Brieskorn, E.: Die Auflösung der rationalen Singularitäten holomorpher Abbildungen; Math. Ann. 178, 255-270 (1968)

[13] Brieskorn, E.: Singular elements of semisimple algebraic groups; Actes Cong. Int. Math. Nice 1970, t.2, 279-284

[14] Brieskorn, E.: Die Monodromie der isolierten Singularitäten von Hyperflächen; Manuscripta math. 2, 103-161 (1970)

[15] Brieskorn, E.: The Development of Geometry and Topology, Notes of introductory lectures given at the University of La Habana, 1973; Mat. z. Berufspraxis Math. 17, 109-203 (1976)

[16] Cayley, A.: A memoir on cubic surfaces; Phil. Trans. Roy. Soc. London 159, 231-326 (1869)

[17] Coxeter, H.S.M., Moser, W.O.J.: Generators and Relations for Discrete Groups, 3rd ed.; Springer, Berlin-Heidelberg-New York, 1975

[18] Demazure, M.: Désingularisation des variétés de Schubert generalisées, Ann. Sc. l'E.N.S. 7, 53-88 (1974)

[19] Drucker, D., Frohardt, D.: Irreducible root systems and finite linear groups of degree two; Bull. London Math. Soc. 14 (2), 142-148 (1982)

[20] Durfee, A.H.: Fifteen characterizations of rational double points and simple critical points; L'Enseignement mathematique, T. XXV 1-2, 131-163 (1979)

[21] Du Val, P.: On isolated singularities of surfaces that do not affect the conditions of adjunction I, II, III; Proc. Cambridge Phil. Soc. 30, 483-491 (1934)

[22] Gabriel, P.: Unzerlegbare Darstellungen I, Manuscripta math. 6, 71-103 (1972)

[23] Gonzalez-Sprinberg, G., Verdier, J.L.: Points doubles rationnels et représentations de groupes; C.R. Acad. Sc. Paris 293, 111-113 (1981)

[24] Gonzalez-Sprinberg, G., Verdier, J.L.: Construction géométrique de la correspondence de McKay; Ann. Sc. E.N.S., to appear

[25] Grauert, H.: Über die Deformationen isolierter Singularitäten analytischer Mengen; Inventiones math. 15, 171-198 (1972)

[26] Happel, D., Preiser, U., Ringel, C.M.: Binary polyhedral groups and Euclidean diagrams; Manuscripta math. 31, 317-329 (1980)

[27] Hirzebruch, F.: The topology of normal singularities of an algebraic surface; Sém. Bourbaki No. 250, 1962-63

[28] Humphreys: Linear Algebraic Groups, Springer Verlag, Berlin-Heidelberg-New York, 1975

[29] Husein-Zade: Intersection matrices for certain singularities of functions of two variables; Funct. Anal. Appl. 8, 11-15 (1974)

[30] Kac, V.G.: Infinite root systems, representations of graphs, and invariant theory; Inventiones math. 56, 57-92 (1980)

[31] Kas, A., Schlessinger, M.: On the versal deformation of a complex space with an isolated singularity; Math. Ann. 196, 23-29 (1972)

[32] Kepler, J.: Mysterium Cosmographicum; Tübingen 1596

[33] Kirby, D.: The structure of an isolated multiple point of a surface I, II, III; Proc. London Math. Soc.(3) 6, 597-609 (1956), 7, 1-28 (1957)

[34] Klein, F.: Vorlesungen über das Ikosaeder und die Auflösung der Gleichungen vom fünften Grade; Teubner, Leipzig 1884

[35] Knörrer, H.: Group representations and resolution of rational double points, preprint Bonn University

[36] Kodaira, K.: On compact complex analytic surfaces II; Annals of Math. 77, 563-626 (1963)

[37] Kostant, B.: Lie group representations on polynomial rings; Amer. J. Math. 85, 327-404 (1963)

[38] Lipman, J.: Rational singularities, with applications to algebraic surfaces and unique factorization; Publ. Math. IHES 36, 195-279 (1969)

[39] Looijenga, E.: On the semi-universal deformation of a simple elliptic singularity II, Topology 17, 23-40 (1978)

[40] Looijenga, E.: Rational surfaces with an anticanonical cycle; Annals of Math. 114, 267-322 (1981)

[41] Mather, J.: Stability of C^∞-mappings I-VI, Annals of Math. 87, 89-104 (1968), 89, 254-291 (1969), Publ. Math. IHES 35, 127-156 (1968), 37, 223-248 (1969), Advances in Math. 4, 301-336 (1970), Lecture Notes in Math. 192, 207-253 (1971)

[42] McKay, J.: Graphs, singularities, and finite groups; Proc. Symp. Pure Math. 37, 183-186 (1980)

[43] Milnor, J.: Singular points of complex hypersurfaces, Annals of Math. Study 61, 1968

[44] Mumford, D.: The topology of normal singularities of an algebraic surface and a criterion for simplicity; Publ. Math. IHES 9, 5-22 (1961)

[45] Pham, F.: Formules de Picard-Lefschetz généralisées et ramification des integrales; Bull. Soc. Math. France 93, 333-367 (1965)

[46] Pinkham, H.: Singularités rationelles des surfaces, Séminaire sur les singularités des surfaces, Lecture Notes in Math. 777, 147-178 (1980)

[47] Plato: Timaios

[48] Schläfli, L.: On the distribution of surfaces of the third order into species; Phil. Trans. Roy. Soc. London 153, 193-241 (1863)

[49] Schwarz, H.A.: Über diejenigen Fälle, in welchen die Gaussische hypergeometrische Reihe eine algebraische Funktion ihres vierten Elementes darstellt; J. Reine Angew. Math. 75, 292-335 (1873)

[50] Sekiguchi, J., Shimizu, Y.: Simple singularities and infinitesimally symmetric spaces, Proc. Japan Acad. 57(A), 42-46 (1981)

[50'] Sekiguchi, J.: The nilpotent subvariety of the vector space associated to a symmetric pair; preprint, Tokyo Metropolitan University, 1983

[51] Serre, J.P.: Représentations lineaires des groupes finis, Hermann, Paris 1967

[52] Slodowy, P.: Simple singularities and simple algebraic groups; Lecture Notes in Math. 815, 1980

[53] Slodowy, P.: Four lectures on simple groups and singularities; Communications of the Mathematical Institute 11, Rijksuniversiteit Utrecht, 1980

[54] Slodowy, P.: Chevalley groups over $\mathbb{C}((t))$ and deformations of simply elliptic singularities. RIMS Kokyuroku 415, 19-38 (1981), Kyoto University

[55] Slodowy, P.: Sur les groupes finis attachés aux singularités simples, Séminaire Lê, Université de Paris VII, 26 mars '81

[56] Slodowy, P.: A character approach to Looijenga's invariant theory for generalized root systems, preprint, Bonn University 1982

[57] Springer, T.A.: Invariant theory; Lecture Notes in Math. 585, 1977

[58] Steinberg, R.: Conjugacy classes in algebraic groups; Lecture Notes in Math 366, 1974

[59] Steinberg, R.: Kleinian singularities and unipotent elements; Proc. Symp. Pure Math. 37, 265-270 (1980)

[60] Steinberg, R.: Subgroups of SU_2 and Dynkin diagrams; Preprint UCLA, 1981

[61] Thom, R.: Les singularités des applications differentiables, Ann. Inst. Fourier 6, 43-87 (1955)

[62] Thom, R.: Stabilité Structurelle et Morphogénèse, Benjamin, Reading Mass., 1972

[63] Tjurina, G.N.: Locally semiuniversal flat deformations of isolated singularities of complex spaces, Math. USSR Izvestija 3(5), 967-999 (1969)

[64] Tjurina, G.N.: Resolutions of singularities of flat deformations of rational double points, Funct. Anal. Appl. 4(1), 68-73 (1970)

[65] Waterhouse, W.C.: The discovery of the regular solids; Arch. Hist. Ex. Sci. 9, 212-221 (1972)

[66] Weyl, H.: Symmetry; Princeton Univ. Press, Princeton N.J. 1952

[67] Whitney, H.: On singularities of mappings of Euclidean spaces I, Ann. of Math. 62, 374-410 (1955).